The Cognitive Continuum of Electronic Music

The Cognitive Continuum of Electronic Music

Anıl Çamcı

BLOOMSBURY ACADEMIC
NEW YORK • LONDON • OXFORD • NEW DELHI • SYDNEY

BLOOMSBURY ACADEMIC
Bloomsbury Publishing Inc
1385 Broadway, New York, NY 10018, USA
50 Bedford Square, London, WC1B 3DP, UK
29 Earlsfort Terrace, Dublin 2, Ireland

BLOOMSBURY, BLOOMSBURY ACADEMIC and the Diana logo are trademarks of Bloomsbury Publishing Plc

First published in the United States of America 2022
This paperback edition published 2022

Copyright © Anıl Çamcı, 2022

For legal purposes the Acknowledgments on p.175 constitute an extension of this copyright page.

Cover design: Louise Dugdale
Cover image: Anıl Çamcı

All rights reserved. No part of this publication may be reproduced or transmitted in any form or by any means, electronic or mechanical, including photocopying, recording, or any information storage or retrieval system, without prior permission in writing from the publishers.

Bloomsbury Publishing Inc does not have any control over, or responsibility for, any third-party websites referred to or in this book. All internet addresses given in this book were correct at the time of going to press. The author and publisher regret any inconvenience caused if addresses have changed or sites have ceased to exist, but can accept no responsibility for any such changes.

A catalog record for this book is available from the Library of Congress.

ISBN: HB: 978-1-5013-5712-1
PB: 978-1-5013-8807-1
ePDF: 978-1-5013-5714-5
eBook: 978-1-5013-5713-8

Typeset by Newgen KnowledgeWorks Pvt. Ltd., Chennai, India

To find out more about our authors and books visit www.bloomsbury.com and sign up for our newsletters.

Contents

List of Figures	ix
Introduction	1
Overview of the Book	4
1 Defining Electronic Music	7
Early Years of Electronic Music	8
The Electronic Medium	8
Nonmusical Sound	9
Birth of the Electronic Music Studio	11
A New Frontier	13
Electronic Music Nomenclature	14
The Medium Is the Genre	15
Acousmatic Music	16
Electroacoustic Music	17
Broadening Horizons	17
2 Situating the Electronic Music Experience	23
Foundations of Musical Behavior	23
Evolutionary Perspectives	24
From Biology to Culture	26
The Material and Language of Music	27
Music and Emotion	31
Affect in Music	33
Experiential Idiosyncrasies of Electronic Music	34
The Composer, Who Is Also a Listener	35
From Parameters to Instincts	40
Complexity of Listening	43
An Amalgamation of Languages	47
Threads of Communication in Electronic Music	48
The Poietic Thread	49
The Esthesic Thread	52

3	A Study on Listening Imagination	55
	A Cognitive Approach	55
	Experimental Studies on Electronic Music	56
	"Talking about Music Is Like …"	61
	Stimuli	62
	Birdfish (2012, 4′40″)	65
	Sound Design	65
	Form	67
	Element Yon (2011, 3′45″)	69
	Sound Design	70
	Form	70
	Christmas 2013 (2011, 2′16″)	71
	Sound Design	72
	Form	73
	Diegese (2013, 1′54″)	74
	Sound Design	75
	Form	77
	Touche pas (by Curtis Roads, 2009, 5′30″)	78
	Study Design	79
	Preliminary Studies	80
	Participants	81
	Setup	82
	Procedure	82
	Initial Listening Session	82
	General Impressions Task	83
	Real-Time Descriptors Exercise	83
	Real-Time Descriptors Task	85
	Results	86
	Data Visualizations	86
	Single-Timeline Dynamic Visualization	87
	Multiple-Timeline Visualization	88
	Analysis Methods	88
	Categorization of Descriptors	89
	Comparative Analysis	92
	Correspondence Analysis	93
	Discourse Analysis	93

4	The Electronic Gesture	99
	Events in the Environment	99
	Environmental Sounds	100
	Models of Mental Representation	102
	Affordances	104
	Gestures in Electronic Music	106
	A Gesture in Electronic Music …	109
	… Is a Meaningful Narrative Unit …	109
	… Operates within Causal Networks …	112
	… Coexists with Other Gestures in Various Temporal and Spatial Configurations …	115
	… Implies Intentionality	117
5	Worldmaking in Electronic Music	121
	Diegesis	122
	An Interdisciplinary Contextualization of Diegesis	123
	Coalescence of Mimesis and Diegesis	125
	Presentationality	126
	Narrativity	127
	Diegetic Affordances and Affect	132
	Music as a Diegetic Actor	137
	Quoting Music within Music	138
	A Diegetic Actor as Music: Electronic Music and Science Fiction	141
6	Tracing the Continuum	143
	Domains of Experience	143
	The Physical Domain	144
	Awareness of the Physical Self	144
	Stream Segregation	146
	Habituation	147
	The Semantic Domain	150
	Effects of Semantic Context	150
	Semantic Gestalts	153
	Signs of Life	154
	Contacts between the Two Domains	155
	Inside and Outside the Diegesis	155

Sense of Time	157
Experienced Listeners	158
Presence of the Composer in the Work	159
Case Study: *Little Animals*	160
Macroscale Analysis	161
Gestural Layers	162
Organic and Environmental Sounds	163
Physical Causalities	164
Pitched and Droning Elements	164
Temporal Flow	165
Diegetic Disposition of the Listener	166
Coda	171
Acknowledgments	175
Bibliography	177
Index	193

Figures

2.1	Digitized excerpts from my notations for two pieces, *Diegese* and *Christmas 2013*	36
3.1	Simple bird-vocalization synthesis using an envelope generator and two oscillators	66
3.2	Overall structure of *Birdfish* shown on a waveform display of the piece	68
3.3	Overall structure of *Element Yon* shown on a waveform display of the piece	70
3.4	Overall structure of *Christmas 2013* shown on a waveform display of the piece	73
3.5	Overall structure of *Diegese* shown on a waveform display of the piece	77
3.6	Initial Listening Session software interface	83
3.7	Participant information form	84
3.8	Real-Time Descriptors Exercise software interface	84
3.9	Real-Time Descriptors Task software interface	85
3.10	Single-timeline dynamic visualization of real-time descriptors	87
3.11	Multiple-timeline visualization of real-time descriptors by two participants	88
3.12	Categorization of the descriptors gathered from the second preliminary study	89
3.13	Categorical distribution of real-time descriptors by piece	92
3.14	Correspondence analysis between pieces and descriptor categories	94
4.1	Overall categorical distribution of real-time descriptors	111
5.1	A participant's general impression of *Birdfish* in the form of a drawing	131
5.2	Another general impression expressed in drawing in response to *Touche pas*	136
6.1	Overall structure of *Little Animals* shown on a waveform display of the piece	161

Introduction

Electronic music is a powerful form of artistic expression for a number of reasons. Emerging from a strong collaboration between music and technology, it leverages the creative potential of the electronic medium and allows any audible sound to be contextualized as musical material. This expansion of material opens up structural possibilities as spectrum, dynamics, space, and time become continuous dimensions of articulation on various scales of musical form. The materials and structures introduced into music through the electronic medium fundamentally challenge our notions of musical meaning. What we hear in electronic music can venture beyond what we traditionally characterize as musical sound and match the auditory complexity of the sounds we encounter in our daily environments. This way, electronic music engages with listening abilities that we take for granted in our everyday lives and shows us how intricate they can be. It can test the boundaries of not only our auditory perception but also our imagination. It can make us envision realities separate from the one we inhabit. It can also make us conscious of our physical environment and our embodied presence in it. Such affordances of electronic music amount to unique experiential qualities for composers and listeners alike.

In 1972, the composer Daphne Oram remarked that the electronic medium made it possible for composers to gain direct control over sounds without needing to put their musical ideas into notation and have them interpreted by a performer. This unmediated ability to manipulate a vast vocabulary of sounds opened up new creative opportunities for composers. Yet, many of those opportunities were not immediately obvious or easily accessible. In 1977, the composer Pierre Boulez wrote that musical invention had been restrained by a divide between the conception and the realization of creative ideas; he viewed it incumbent upon artists to bridge this gap by furthering their comprehension of contemporary technology. Over the ensuing decades, composers have indeed firmed their grasp of the electronic medium, which itself has become more amenable to artistic use with the introduction of new audio technologies. However, composers' increased access to technology brought about new challenges, this time, of aesthetic nature. In 1997, the composer Denis Smalley observed that one of the major hurdles that composers faced was to maintain an aesthetic path in the "wide-open sound world" of electronic music. Since then, composers' dominion over sound has only expanded as computational systems grew increasingly capable of executing complex audio processes in real time. Today, the electronic medium affords a practically instantaneous dialogue between creative actions and their audible outcomes, allowing composers to imagine through the endless possibilities of sound.

If listening to music can be broadly characterized as an aesthetic experience of contrasts and surprises across various dimensions of sound, composition could be viewed as an act of building up auditory expectations, and then either meeting or evading them. But if anything can be expected from the wide-open sound world of electronic music, how can it evoke a musical sense of anticipation? In this book, I argue that the network of expectations in electronic music is substantially informed by our everyday lives. This is not to say that all composers draw material or inspiration from their environments. Neither do I claim that listening to electronic music is rooted exclusively in representations. But, as I discuss in Chapter 2, abstractness stems from a negation of reality. As we parse through our experience of an electronic music piece, our existing notions of reality form a basis for our interpretation of unreality. I argue that when the virtually unlimited vocabulary of electronic music expands that of a culturally established language of music, it instigates for the listener a profusion of references rooted in events in the environment. I specify events as the units by which perceived time moves forward. Based on existing studies on auditory perception and models of mental representation, I build a thread across events, environmental sounds, and electronic music. This thread leads us to an idiomatic definition of gesture as a meaningful and intentional narrative unit in electronic music. With that, I also stress how the meaning and intentionality conveyed through sound can be as much the listener's construction as it is the product of a composer's expression--if such an expression exists in the first place. This understanding liberates the electronic gesture from a communicational hierarchy and places the emphasis on a complexity of listening fostered by both the composer and the listener.

I expand on this view by characterizing listeners' engagement with electronic music as an act of worldmaking. As listeners construct meaning from their experience of a piece, they effectively superimpose a semantic space on top of the physical reality of their listening environment. Using examples from listener feedback, I delineate both conceptual and perceptual relationships between the semantic and the physical domains of the listening experience. I adopt the narratological concept of diegesis to articulate how listeners contextualize themselves and the composer in relation to the implied universe of a piece, and how fluid this contextualization can become. Listeners not only populate this universe with objects that are appropriate to their narrative constructions but also assume shifting perspectives toward it over the course of a piece. They can, for instance, position themselves as part of the narrative and adopt a first-person view. This can then transform into a third-person view, where they assume the role of an onlooker observing the unfolding of a series of events. To articulate the narrative disposition of the listener with respect to both the perceptual and the semantic affordances of electronic music, I describe a coalescence of narrative modes informed by cross-disciplinary interpretations of diegesis and situate electronic music in a broad framework of artistic disciplines, including literature, film, and visual arts.

For this book, I conducted a listening study with eighty subjects over the course of four years to gain a deeper understanding of how fixed pieces of electronic music operate on perceptual, cognitive, and affective levels. The study was designed to capture a detailed account of a listener's experience with electronic music by gathering both immediate and post-hoc impressions that reflect how they conceptualized this experience. These impressions offer insights into the communication of meaning in electronic music and how concepts and narratives drawn from a piece can overlap and diverge not only among listeners but also between a composer and their audience. To make sense of the vast amount of feedback gleaned from the study, I employ a diverse range of analysis techniques, including data visualization, descriptor categorization, and discourse analysis, among others. Throughout the book, I refer to the results of this study to weave links between artistic practice, cognitive psychology, linguistics, and philosophy. In doing so, I delineate a cognitive continuum as an intrinsic aspect of the electronic music experience. This continuum spans from abstract to representational based on the relationship of gestures in electronic music to events in the environment, guiding our appreciation of this music through our past encounters with auditory phenomena.

Ultimately, this book is a testament to the depth and breadth of experiences that electronic music can evoke. The dichotomy between representationality and abstractness has historically been a cause of debate not only in electronic music but also in the arts more generally. In this book, I shift this conversation from a binary distinction between the two toward a continuum, wherein the cognitive reciprocity between the representational and the abstract traits of what we take away from a piece of electronic music underlies the richness and intricacy of our experience with it. In 1986, the composer Simon Emmerson argued that even if a composer is not interested in manipulating the images associated with electronic music, they must take into account the duality between mimetic and aural aspects of the listening experience. In the same article, Emmerson called on future researchers to combine psychology of music with analyses of symbolic representation and communication at a deeper level to explore what makes particular sound combinations in electronic music work. I believe that this book addresses this appeal in its pursuit to further our awareness and understanding of the cognitive continuum of electronic music.

Overview of the Book

In Chapter 1, "Defining Electronic Music," we travel back to the early years of the twentieth century to examine the birth of electronic music amid new philosophies of sound, on the one hand, and emerging technologies, on the other. I give examples of how experiments in the electronic medium brought about new techniques that in some cases became assimilated into existing styles whereas in other cases prompted all-new styles or experimental traditions. As we find our way back to the modern day, we also get a sense of the book's aesthetic framework.

In Chapter 2, "Situating the Electronic Music Experience," we explore some of the experiential idiosyncrasies of electronic music. First, I survey the evolutionary origins of musical behavior, discussing the biological and cultural factors that govern our appreciation of music. Then, I narrow this discussion down to electronic music to identify the unique aspects of how it's created and experienced. In dealing with the communication of meaning in electronic music, I utilize Molino's model of semiology to emphasize the listener's creative license in constructing meaning from their experience. At the same time, I build up on the historical overview presented in Chapter 1 as I discuss the evolution of the relationship between artists and technology through the modern day.

In Chapter 3, "A Study on Listening Imagination," I introduce one of the main contributions of this book: a listening study that investigates the cognitive foundations of our engagement with electronic music. I offer an extensive discussion of the creative intents, tools, and techniques underlying the five works of electronic music used in this study. I then describe the novel experimental design of the study and the analysis techniques used to interpret its outcome. The results presented in this chapter already give us some unique insights, for instance, into which general categories can be gleaned from multivariate listener impressions. These results inform the theoretical discussions throughout the rest of the book.

In Chapter 4, "The Electronic Gesture," I formulate a relationship between our everyday environments and electronic music at the level of events that function as structural units. I begin by mapping out a phenomenology of environmental events by surveying models of mental representation that are conceived to explain how we deal with external stimuli in our everyday environments. To articulate the unitary role of events in how we parse electronic music, I focus on the concept of gesture and its numerous interpretations across disciplines. I then utilize the study results to devise an idiomatic definition of the electronic gesture as an intentional narrative unit that coexists with other gestures in various temporal and spatial configurations, and within causal networks.

In Chapter 5, "Worldmaking in Electronic Music," I expand the scope of our discussion from gestures to the broader contexts they establish on various scales of form. I adopt Goodman's theory of worldmaking to portray how listeners construct narratives as they imbue their experience of an electronic music piece with meaning and intentionality. This opens up a broader discussion of narrativity in electronic music where we explore how the elements of a listener's narrative are situated inside or outside the spatiotemporal universe (i.e., the diegesis) that a piece might evoke. I introduce the concept of diegetic affordances to describe listeners' encounters with the dimensions, landscapes, surfaces, and objects of such a universe.

In Chapter 6, "Tracing the Continuum," I expand upon the worldmaking approach to identify some of the explicit and implicit aspects of a listener's construction of a narrative. After I articulate the physical and the semantic domains wherein such narratives are grounded, I discuss the ways in which these two domains can come in contact and how such contacts can actively shape the listener's interpretation of a piece. I then analyze Natasha Barrett's piece *Little Animals* in this framework, using some of the theoretical constructs presented so far.

1

Defining Electronic Music

What is electronic music? The answer to this question can be deceivingly simple: it's a kind of music made by electronic means. The word *electronic*, however, is both technologically and historically loaded. The modern use of this word dates back to the invention of the vacuum tube in the early twentieth century. The vacuum tube, which is a component that controls the flow of electric current, was used as a building block in early electronic systems, including the first-generation computers. The introduction of the transistor in the 1940s facilitated the design of more efficient and compact systems, paving the way for consumer electronics all the way through the modern day with personal and mobile computers. The field of electronics therefore encompasses a wide range of technologies spanning multiple eras. Today, the word *electronic* can be used in reference to electrons, electronic components, and digital systems such as networks and computers. When used in the context of music, this broad concept conjures up many practices that involve the use of computers, synthesizers, and other electrical or electromechanical means to generate musical sound. The stylistic scope of electronic music is also broad. The evolution of the field of electronics throughout the twentieth century is mirrored in electronic music with new technologies prompting new musical styles and practices. Many modern music genres, such as electronica, techno, house, and ambient, are colloquially referred to as electronic music. So, what does electronic music mean as it applies to its use in this book? What are the aesthetic, technical, and conceptual underpinnings of using this term? How do such considerations inform the artistic works and the theories presented throughout the book? To answer these questions, we will start off with a brief historical overview of electronic music. In doing so, we will also begin to find out what makes it so unique as a form of creative expression.

Early Years of Electronic Music

There are many evolutionary and cultural factors that motivate us to engage in musical behavior. We will explore some of these factors more explicitly in the next chapter. But the ideas that propelled the early days of electronic music can be attributed to humankind's perennial obsession with making sounds. It is therefore difficult to pin the conceptual genesis of electronic music on a person or event. We can, however, trace its technological origins to better understand the emergence of the electronic medium and the musical paradigm shifts it brought about.

The Electronic Medium

The earliest technology that allowed the electrical transmission of sound is the telephone. By the 1860s, many inventors had been experimenting with electrical telephony based on the transmission of signals between two transducers: a microphone, which converts acoustic waves into electrical signals, and a loudspeaker, which converts these electrical signals back into acoustic waves. In combination, these two devices have ushered sound into the electronic medium. Among the fields where telephony would find use, music trailed telecommunication as a foremost domain of application. Philipp Reis, one of the first scientists to give a successful demonstration of electrical telephony, had actually envisioned his invention as a music broadcasting system, therefore calling it the "singing station" (Tucker 1976).

In 1857, Édouard-Léon Scott de Martinville introduced the phonautograph, a device that etched acoustic waves onto paper. Although the phonautograph was not capable of playing back these etchings, it nevertheless produced the first recordings of acoustic signals (Feaster 2019). Soon after, Thomas Edison introduced the phonograph, which relied on a similar principle for capturing sounds but with the added ability to play back its recordings. Although the earliest iteration of this technology pales in comparison to modern audio tools in terms of sound fidelity, it nevertheless made it possible for sounds to be detached from their physical sources and stored for playback at a later time. With the invention of the vacuum tube at the turn of the twentieth century, new techniques for sound generation and amplification became available. This has not only enabled higher-quality audio reproduction systems but also facilitated the design of electronic music instruments such as the theremin, ondes Martenot, and Trautonium.

It was not long after such advances in electronics that the artistic implications of the electronic medium were recognized. Musical sound could now be decoupled from the acoustic and ergonomic constraints of traditional music performance: Sounds could be amplified well beyond the levels achievable with an acoustic instrument played by a human performer. Sounds could also be frozen in time, sped up, slowed down, reversed, and layered in ways that were previously impossible. While some viewed the electronic medium as a means to rid music of human error (Grainger 1996), others used it to explore entirely new paradigms for music composition. In the early 1930s, the composers Paul Hindemith and Ernst Toch were among the first to exploit sound recording and reproduction techniques to layer and juxtapose musical passages. They would play recordings of these passages at varying speeds and in different directions to create unique timbres and textures. This style of music-making, known as *Grammophonmusik*, is considered among the precursors to modern turntablism (Katz 2001). In 1939, the composer John Cage introduced *Imaginary Landscape No. 1*, one of the first pieces to combine acoustic instruments with live electronic sound. Cage used two phonographs to play back sine waves at variable rates to create uninterrupted tone glides that accompanied prepared piano and cymbal. Around the same time, the composer Halim El-Dabh began to experiment with the wire recorder. This technology was the first implementation of magnetic recording, which gained broader appeal with the introduction of the tape machine. El-Dabh's experiments with the wire recorder paved the way for his 1944 composition *The Expression of Zaar*, which is one of the earliest works of tape music.

Nonmusical Sound

In the early twentieth century, the preconceived notions of material in art were being challenged across numerous artistic disciplines. Although many classical composers had already been experimenting with new timbres, scales, and rhythmic structures, the tenets of avant-garde art movements such as Dadaism and expressionism began to resonate in musical works. The widespread industrialization that preceded the twentieth century and the growing international tension leading up to the First World War had deep societal impacts that invigorated some of these movements.

In 1913, the Futurist painter and composer Luigi Russolo published a manifesto where he laid out his vision of how music would evolve in a world that was going through major upheaval. Reflecting the Futurist fascination with technology, Russolo formulated an "art of noises" rooted in the proliferation of machinery, arguing that the audiences would eventually develop a taste for the infinite variety of noises brought about by the machines. Although we have yet to find the engines of our industrialized cities skillfully tuned into "an intoxicating orchestra of noises" as Russolo foresaw, he accurately depicted a need for expanding the domain of sounds that were deemed to be musical (Russolo 1967: 11). In 1916, the composer Edgard Varèse made a similar plea in an interview with the *New York Morning Telegraph*, where he expressed a need for our musical alphabet to be enriched (Chadabe 1997: 59). Although a similar sentiment had already begun to find a foothold in classical music (Samson 1977), Russolo and Varèse both foregrounded the role of nonmusical sounds in materializing this vision.

But what exactly is a nonmusical sound? The negation of musical sound was a criticism of the status quo in music back then. According to Varèse, the prevailing alphabet of musical sounds was "poor and illogical" (Varèse and Wenchung 1966: 11). He viewed electronic sound as the most promising means to address this problem. Besides promoting the potential of electronic sound in his writings, Varèse also petitioned funding agencies to support the development of electronic music instruments.[1] Highlighting its virtually limitless possibilities for musical expression, he argued that the electronic medium would open up for composers a "whole mysterious world of sound" (18).

In his hypothesis on the origins of music, Russolo associates humankind's initial fascination with musical instruments to the prevalent quietness of life in early human history (1967: 4). He suggests that this fascination conditioned people to attribute a divine origin to musical sound as a phenomenon decoupled from everyday life and prompted a perception of music as an "inviolable and sacred world" superimposed upon reality (5). Russolo then extrapolates this theory to the modern world and argues that the proliferation of machinery in the nineteenth century had already begun to challenge this fantasy, mandating a revolution of music wherein we would "break at all cost from [the] restrictive circle of pure sounds and conquer the infinite variety of noise-sounds" (6).

---

[1] During the first half of the twentieth century, Varèse made several appeals to organizations like the Guggenheim Foundation, Bell Laboratories, and various production studios in Hollywood to urge them to invest into electronic music technologies (Chadabe 1997: 59; Doornbusch 2011). Although these attempts were unsuccessful at the time, the said organizations would become major proponents of electronic music in the second half of the century.

His suggested means of achieving this was the *intonarumori* (noise intoners). These were mechanical devices designed to produce a specific set of nonmusical sounds categorized into groups of what Russolo referred to as "fundamental noises" including roars, whistles, cracks, and bellows. The sacrilege of musical sound was to be perpetrated by a futurist orchestra of noise intoners.

In 1937, John Cage expressed a similar view when he said, "If the word 'music' is sacred and reserved for eighteenth- and nineteenth-century instruments, we can substitute a more meaningful term: organization of sound" (Cage 2004: 26). In this proposal, Cage echoes the terminology of Varèse, who is credited for coining the term *organized sound* in the 1920s. During a lecture at Yale University in 1962, Varèse gave an account of why he preferred this term:

> Although [electronic music] is being gradually accepted, there are still people who, while admitting that it is "interesting," say, "but is it music?" It is a question I am only too familiar with. Until quite recently I used to hear it so often in regard to my own works, that, as far back as the twenties, I decided to call my music "organized sound" and myself, not a musician, but "a worker in rhythms, frequencies, and intensities." Indeed, to stubbornly conditioned ears, anything new in music has always been called noise. But after all what is music but organized noises? (Varèse and Wen-chung 1966: 18)

Here, Varèse initially decouples his work from a generalized notion of music but promptly acknowledges that the seemingly eccentric interpretation of his practice as organized sound could in fact apply to all music. This was neither the first nor the last time the categorization of a new art form would be problematized. Music was evolving into something new, as it always did. But could it still be called music if it broke away from tradition altogether? Electronic music both sprouted from and fostered a liberation from musical dogmas.

Birth of the Electronic Music Studio

Halfway into the twentieth century, some of the earlier predictions about the future of music started taking shape. Technology was becoming entrenched into not only how music was created but also how it was distributed and consumed. Public broadcasting agencies around the world began to take an interest in electronic music as a way to put these technologies to test. In a postwar Europe, the first electronic music studios were propped up with phonographs, magnetic tape machines, reverberators, filters, and oscillators. These institutions would soon explore the fringes of new music at the intersection of art and science. The interlacing of music and research, as well as a perpetual interplay between aesthetic persuasions and technological advances, gave rise to a proliferation of styles in electronic music.

Radiodiffusion Française was among the first state institutions to provide support for research into electronic sound. Starting in the 1940s under the leadership of the composer and broadcaster Pierre Schaeffer, the Paris studio would host many artists and researchers who leveraged phonographs and tape machines to explore the manipulation of recorded sound. The aesthetic style of this group, which would eventually be named Groupe de Recherches Musicale, was known as *musique concrète*. Formulated by Schaeffer as "a new mental framework for composing" (Schmidt 1981), this style was characterized by its manipulation of tape recordings into assemblages of sound objects that demand a reduction of the listening experience, where a sound recording would be perceived as a phenomenon in and of itself rather than an auditory indicator of another object or event (Kane 2007). A sense of play and open-ended exploration was central to this plasticization of music by way of abstracting sounds from their sources.

In the early 1950s, another early electronic music studio was established in Germany at the Westdeutscher Rundfunk in Cologne. The stylistic direction of this studio was drawn from the twelve-tone serialism of the Second Viennese School. The twelve-tone technique utilizes tone rows wherein any one of the twelve pitches common to the Western musical tradition can only be utilized once, ensuring the absence of a tonal center in the resulting music. Extending this procedural technique to the serialization of other audio parameters, the Cologne studio adopted *total serialism* as its stylistic guideline. Oscillators and noise generators, which facilitated the kind of parametric precision inherent to this style, were the primary sound sources for the composers of the Cologne studio. Their artistic output was simply referred to as *elektronische Musik*.

The sense of play inherent to musique concrète stood in stark contrast with the absolute determinacy of total serialism, and some composers were unabashed in expressing their views on this aesthetic conflict. For Herbert Eimert, who was one of the founding members of the Cologne studio, musique concrète was "fashionable and surrealistic" and consisted of "incidental manipulations or distortions haphazardly put together" (Holmes 2008: 58). In return, serialism was criticized by prominent members of the Paris studio: according to the composer

Iannis Xenakis (1992), the complexity inherent to serial music amounted to "an irrational and fortuitous dispersion of sounds" that the listeners were unable to follow.

A New Frontier

In retrospect, these distinctive approaches propelled the early days of electronic music and shepherded many composers into aesthetically uncharted territory. The stylistic principles adopted by the studios guided composers, most of whom came from a background in pen-and-paper composition, through their first encounters with the electronic medium. *Études* of musique concrète or *studies* of elektronische Musik were, to the most part, cautious stabs at an entirely new form of artistic expression, mirroring the distinct aesthetic inclinations of the studios. But at the same time, these exercises granted the composers a much-needed technical sensibility. Those who envisioned a long-standing relationship with electronic music were quick to integrate the unique implications of this medium into their craft. It was not long before artistic instincts began to overcome the predetermined credos of the studios. The stylistic barriers between the seemingly segregated schools of practice would dissolve in a matter of a few years. In her discussion of this stylistic evolution in electronic music throughout the 1950s, the composer Daphne Oram remarks how the greatest music was composed when a composer gained "sufficient strength of character to control his forces by his own individuality" (1972: 39).

In the next chapter, we will further investigate how composers began to move beyond stylistic boundaries as early as the 1950s. However, this very evolution marks the emergence of the musical style this book will focus on: a style of music that leverages electronic sound to give precedence to the composer's vision and instincts; a style that thrives on the composer's ability to imagine through the endless possibilities of sound; a style where the creative constraints and opportunities are intrinsically tied to the electronic medium. Ultimately, this medium governs both the structure and the sound of the music that is made with it (Barrett 2013, personal communication). The composer Morton Subotnick reflects a similar view when he describes his first contemplations of electronic music in the early 1960s:

> I didn't know what kind of music it would be. I did believe that it should be a music for that medium, not just any old piece of music that got recorded

and put out—a whole new kind of music. I used as a model for myself the metaphor of Chopin and the piano. He used the concert grand in a way that just wouldn't have made sense as anything but solo piano pieces. It was (media theorist Marshall) McLuhan's idea that "the medium is the message," that his piano music and the piano were linked together in a very special way. I was looking for something equivalent with a record. What would a record be if (the piece) never got performed on the stage, just for the record? (Rosenbloom 2011)

As we will explore throughout this book, the interactions between a composer and their medium of composition, whether this involves a piece of staff paper, an instrument, or a computer, has a fundamental influence on the inherent qualities of the work they create in that medium. The reciprocation between a compositional plan and its materialization impacts upon creative strategies on various scales from gesture to form. This reciprocation, which the next chapter will delve deeper into, manifests itself in electronic music as a real-time feedback loop between actions and perceptions (Vaggione 2001: 60). The result is a music that is "fundamentally different in character and in aspiration from any music that preceded it" (Demers 2010: 12), a music that "opens access to all sounds, a bewildering sonic array ranging from the real to the surreal and beyond" (Smalley 1997: 107).

Electronic Music Nomenclature

The technological limits of an emerging medium influences the musical nomenclature associated with it (Spiegel 1992). When the early styles of electronic music were dubbed with names that are indicative of the studios they originated from, there were clear philosophical and technological distinctions that motivated these names. But as the stylistic boundaries between the so-called schools of electronic music began to blur, the etymological pertinence of these names started to degrade with them. To illustrate this, let's examine the current electronic music nomenclature, consisting of both modern and historical phrases that are used—almost interchangeably—to categorize works in this domain. We will discuss the accuracy with which these phrases signify the works to be discussed throughout this book. Needless to say, some of these phrases have gained prominence not merely by virtue of their aptness but also by their widespread usage as idioms. This is why the following discussion is not necessarily intended to disregard the use of these phrases but to further establish a stylistic context for this book.

The Medium Is the Genre

The computer music pioneer Max Matthews's early work on computational music systems at Bell Labs has set us off onto what the composer John Chowning refers to as a "profoundly deep and consequential adventure" (2008: 1). Most of the original music to be discussed in the following chapters can be characterized as *computer music* on the basis that they have been created with a computer. But, following Subotnick's analogy, the computer is no longer an instrument for this music in the sense that a concert grand is an instrument for Chopin's solo piano works. The less a computer is a constraining factor in a compositional workflow, the more useful to the composer it becomes. Computer hardware today is less limiting for composers than it has ever been. For an artist working in the audio domain, the modern computer is a virtually infinite playground outfitted with software tools that can perform complex audio processes in real time.

This was certainly not the case sixty years ago, when computers began to find use in artistic practices. The functional limitations of early computers were entrenched into the work that was produced with them, and the composer was forced to account for these limitations throughout the creative process. For instance, the computers back then could only facilitate a deferred composition workflow with limited processing power; audio code would be programmed onto punch cards to be compiled over hours, if not days. The implications of this workflow justified the naming of its outcome as computer music since the medium had a significant imprint on the creative process. Fortunately, this is no longer the case. Over the past decades, not only have computers become deeply integrated into our daily lives, but the very definition of a computer has become amorphous as mobile, embedded, and ubiquitous computing technologies began to proliferate. Today, most music is produced on computers or computational systems, obviating the need to carve out a style based on this medium.

Another limitation of the phrase *computer music* is that it can be historically exclusive. Most obviously, it excludes any electronic music composed before computers became available for artistic applications. From the listener's perspective, a piece of electronic music composed on tape in the 1960s can display the same experiential qualities as a recent piece of computer music while falling into a different category purely on a taxonomic basis. Furthermore, many electronic music practices today rely on standalone synthesizers, modular systems, or other music technologies that might rely on digital computing but are not traditionally viewed as computers.

On the other end of the historical spectrum, we are reminded of another phrase signified by the medium of composition: tape music. This phrase is still used in certain situations to characterize fixed pieces of electronic music even though the amount of such music produced on tape today is next to none. It holds a nostalgic value as a throwback to the days of tape composition, acknowledging the stylistic continuity between tape music and modern electronic music. However, it suffers from a similar lack of historical comprehensiveness as computer music. Today, the phrase *tape music* is most often used in concert programs to differentiate fixed pieces of electronic music from those that are performed live.

Acousmatic Music

Another common phrase in electronic music nomenclature is *acousmatic music*. In 1960, the poet Jerôme Peignot proposed the word *acousmatic* as a substitute for musique concrète (Peignot 1960), drawing inspiration from Pythagoras, who used it to characterize the lectures he delivered behind a curtain so that his pupils could concentrate on his teachings rather than his presence. Pierre Schaeffer further developed Peignot's proposal and specified acousmatic sound as one that involves taking a natural listening process, wherein sounds would be reflexively associated with their sources, and reducing it down to a mode where sounds would be heard without imagining the causes behind them (Kane 2014). Acousmatic music is therefore defined through a listening attitude rather than the tools and techniques involved in its creation.

Since the 1960s, not only has the meaning of acousmatic music been a cause of debate (McFarlene 2001), but the very possibility of a truly acousmatic experience has been called into question (Smalley 1991). Today, acousmatic music is most commonly used in concert programs, much like tape music, to distinguish fixed pieces from live electronic performances. But since acousmatic sound is in the ear of the beholder, imposing it onto music taxonomically can imply experiential discrepancies. A composer might intend to create a world of extramusical representations, yet the audience can still practice reduced listening and disregard any symbolic meaning. By the same token, a piece of electronic music can be performed live yet still be perceived acousmatically by way of ignoring visual cues. For acousmatic music to function as an inclusive identifier for electronic music, it would need to be divorced from such considerations; yet those are the very considerations that perpetuate the rich philosophy of listening that acousmatic music sprouts from and that justify the use of this concept in the first place.

Electroacoustic Music

Electroacoustic is an acoustic engineering term defined by the *Oxford Dictionary of English* as "involving the direct conversion of electrical into acoustic energy or vice versa" (2012). Based on the concept of energy transduction, the term is associated with the working principles of loudspeakers and microphones. When we treat the phrase as the adjective that it is, electroacoustic music translates to "music that involves the conversion of electrical into acoustic energy, or vice versa," or "music that involves loudspeakers and microphones." However, according to the composer Leigh Landy (2006), the involvement of electricity in electroacoustic music should go beyond a simple use of a microphone or amplification to call it as such. For the composer Per Anders Nilsson (2018), electroacoustic music is a form of "art music" that is created with electronic equipment and heard from loudspeakers. Landy (2006), on the other hand, argues that the style can extend beyond art music to include forms that are rooted in popular music traditions as long as the use of electricity to register or produce sounds is a major focus of the work.

The musicologist Arnold Whittall (2011) characterizes electroacoustic music as a style that combines instrumental sounds with their electronically manipulated versions or other pre-recorded sounds. Similarly, the composer François Bayle (1993) formulated acousmatic music in such a way that the phrase *electroacoustic music* would be reserved for performances involving a mix of live and prerecorded elements. Indeed, electroacoustic music is commonly used today to identify works or performance practices where acoustic sounds are accompanied by live or fixed electronic sounds.

Broadening Horizons

My own works as an electronic music composer have at the same time been referred to as computer music, tape music, acousmatic music, and electroacoustic music. Much like the numerous examples discussed throughout this book, it is a kind of music that leverages the electronic medium to enrich its form and vocabulary. This is why the composer Curtis Roads views electronic music itself as a medium that can serve different styles rather than a musical genre per se, defining it broadly as a "general category of analog and digital technologies, concrète and synthetic sound sources, and systematic and intuitive composition strategies" (2015: x).

Roads argues that the heterogeneity of the sound material in electronic music brings about a heterogeneity of musical structure (6). Indeed, the proliferation of musical material ushered in by the electronic medium has created many musical trends over the past century. Some of this material first emerges in relatively obscure nooks of the musical milieu. Over time, this material might find its way into different styles as the tools and techniques that were once exclusive to avant-garde practices become integrated into more mainstream genres. Furthermore, musical experimentations with a new technique can eventually lead to an adaptation, wherein the said technique, and the new material it brings forth, becomes assimilated into a style. This adaptation not only yields conventions for the use of this new material in the context of the said style but can also prompt further experimentation.

A prominent example of this kind of evolution in electronic music is observed with synthesizers. In 1964, Robert Moog introduced one of the first commercial synthesizers, which incorporated a traditional piano keyboard into its voltage-controlled interaction paradigm. This feature has been regarded as a move away from the novel musical affordances of the synthesizer (Dalgleish 2016). Don Buchla, another prominent designer of synthesizers, had argued that electronic sound was a new source that demanded a new approach to instrument design. He therefore dismissed the idea of introducing the twelve-tone keyboard into his synthesizers (quoted in Pinch and Trocco 1998).

The unique capabilities of the Buchla synthesizer for not only generating but also controlling electronic sound underlie the works of electronic music pioneers such as Morton Subotnick and Suzanne Ciani. Other artists have also supported the abstraction of the synthesizer from a traditional musical interface. According to the composer Hubert Howe, the use of the keyboard brought about a limitation of the spectral and, therefore, the timbral affordances of the synthesizer (1972: 123). For the composer Éliane Radigue, who has composed some of the most influential works to be created with the ARP 2500 synthesizer, abandoning the keyboard in the 1960s was a strategy to distance herself from the musical practices associated with it (Rodgers 2010: 57).

On the other hand, the integration of the piano keyboard in the Moog synthesizer also paved the way for numerous groundbreaking works of electronic music, a prominent example of which is Wendy Carlos's *Switched-On Bach*. In this case, the keyboard interface on the Moog synthesizer served as a bridge between traditional and modern practices in music, drawing people familiar with the Western musical tradition into the world of electronic sound. As a result, the oscillator, once a staple of the Cologne studio, entered the instrumental arsenal of numerous artists operating in a diverse range of musical genres, including those that are mainstream.

Today, while the synthesizer is viewed as a common instrument in popular music, we are also witnessing a resurgence of the modular format, heralding a new wave of experimentation. This time around, the ideas that inspired the early modular designs in the 1960s are augmented with digital signal processing techniques that have been developed since. New modules combine the use of control voltage with the computational power of modern embedded systems to bring novel musical capabilities to modular synthesizers. These contemporary designs find diverse applications in today's music from the ambient textures in the experimental pop band Animal Collective's album *Painting With* to the algorithmic melodies that adorn the composer Caterina Barbieri's album *Patterns of Consciousness*.

A similar evolution can be traced in the use of sampling in music. The early experiments in musique concrète explored the affordances of fixed media, such as vinyl and tape, for manipulating sampled sounds. The BBC Radiophonic Workshop founders Daphne Oram and Desmond Briscoe began to utilize sampling techniques in television and radio plays in the late 1950s, expanding the reach of electronic music to new audiences (Hutton 2003: 50). Around the same time, Xenakis would exploit the use of tape splicing to amalgamate brief audio samples into dense sonic textures; this technique would serve as a foundation for the sampling-based synthesis technique known as granular synthesis.

Throughout the 1960s, sampling found broad applications across numerous musical practices. In 1965's *It's Gonna Rain*, the composer Steve Reich put two tape loops of the same speech recording gradually out of phase to intensify the melody and the meaning of the speech through repetition and rhythm (Reich 1987). A year later, the composer would utilize the same technique in his influential work *Come Out*, which was composed in support of the civil rights movement (Beta 2016). The same year, the composer Pauline Oliveros built an accumulating delay line using two tape machines to create the swelling drones in her piece *I of IV*. The technique that Oliveros developed for this piece inspired many other composers to experiment with tape delays in their work (Holmes 2008: 133).

The use of instrumental tracks in the Jamaican sound system, wherein disc jockeys would play music through stacks of loudspeakers at dance events, has evolved into a musical movement that fostered the birth of deejaying in the 1960s (Howard 2008). The works of the record producers King Tubby and Lee "Scratch" Perry in this domain have spearheaded dub music and more broadly the remixing culture (Partridge 2007). When hip-hop artists such as Kool Herc and Grandmaster Flash, who were heavily influenced by this culture, pioneered the live control of record playback, sampling became a major musical movement that has resonated through numerous genres since the 1970s. Rap musicians have since leveraged sampling as "a means by which the process of repetition and recontextualization can be highlighted and privileged" (Rose 1994: 73).

These examples demonstrate how experiments in the electronic medium yield transformative musical applications that can attract mainstream appeal. Another area wherein electronic music has historically reached broader audiences is visual media. In one of the earlier pairings of film and electronic music, Miklós Rózsa's use of the theremin in Alfred Hitchcock's 1945 movie *Spellbound*, which was awarded an Oscar for best original score, brought recognition to this early electronic instrument as an expressive tool in film music (Hayward 1997: 36). The ethereal sound of the theremin would soon become a staple of film sound, particularly as an accompaniment to suspenseful moments in thrillers.[2]

In another early instance of the film industry turning the spotlight on electronic music, Bebe and Louis Barron used overloaded electronic circuits to create the unique sounds of the 1956 sci-fi film *Forbidden Planet*, exploiting the intuitive link between electronic music and science fiction. A few years later, Delia Derbyshire's electronic theme for the sci-fi series *Doctor Who* would position her as a pioneer of the medium for drawing significant public attention to electronic music (Winter 2015: 15). Less than a decade later, Wendy Carlos applied the groundbreaking style she developed for *Switched-On Bach* to the

---

[2] Interestingly, the same sound would be used in The Beach Boys' 1967 hit *Good Vibrations*, which is arguably one of the most upbeat pop songs in modern history.

soundtrack of Stanley Kubrick's dystopian classic *A Clockwork Orange*, which opens with Carlos's electronic interpretation of Henry Purcell's *Music for the Funeral of Queen Mary*.

These are just a few of the examples that illustrate the unceasing reciprocity between technology and culture in the evolution of electronic music. Within less than a century, we have witnessed electronic music transform our notions of musical sound, form, and performance. Trying to pin down stylistic benchmarks for electronic music can therefore be complicated, if not futile. This is also why drawing distinctions between the experimental genres of the past and the conventional genres of today can prove to be near-sighted. As we have witnessed time and again over the past century, electronic music concepts and techniques sprouting from a small circle of artists and researchers can proliferate in unforeseen ways. As we will find out in the coming chapters, electronic music can evoke in listeners highly complex and deeply personal sensations that may or may not align with the stylistic disposition of the music or the intents and techniques behind it. Throughout this book, I will use "electronic music" as an inclusive term that reflects the technical, aesthetic, and conceptual implications of the electronic medium for musical creativity and the idiosyncratic experiences that it offers to composers and listeners alike. In the next chapter, I will situate these qualities in the broader context of how we engage with musical material. I will expand the historical context presented so far with compositional trends that emerged throughout the second half of the twentieth century. As I extend this discussion into the present day with examples from modern practices, we will gain a broader understanding of the evolving relationship between music and technology.

2

Situating the Electronic Music Experience

In this chapter, we will begin to explore the unique experiential qualities that electronic music offers. To facilitate this conversation, I will first give an overview of our engagement with music in a broader sense, surveying some of the theories on how the human urge to be musical may have emerged in the first place and how our appreciation of music is influenced by both biological and cultural factors. I will then narrow this discussion down to our experiences with electronic music. We will revisit the earlier days of electronic music, this time focusing on the evolution of a priori methods of composition into those that are guided by composers' instincts. I will contextualize "the great opening up of music to all sounds" (Chadabe 1997) and its impact on musical creativity with examples from modern practices. This will lead us into an analysis of the threads of communication in electronic music that are rooted in the complexity of listening.

Foundations of Musical Behavior

A complex web of perceptual and cognitive processes governs our musical experiences, whether these involve the creation or appreciation of music. Despite the significant body of research on the mechanisms through which the human brain parses musical information, many aspects of our affective appraisal of music are yet to be mapped out. In this section, we will go over some of the theories that aim to illuminate why and how we engage in musical activities. These theories will provide a context for our ensuing discussions on how the human mind deals with electronic music in particular.

Evolutionary Perspectives

Since music predates recorded history, we only have hypotheses regarding its origins. Early musical behavior in humans has been linked to a diverse range of functions from communication to ritual. While it is impossible to pinpoint a single instigator as to *why* we started to make music, the *how* is often associated with mimicry. In his book *The Great Animal Orchestra*, the acoustic ecologist Bernard Krause (2012) breaks down our relationship with sounds into several evolutionary phases. His resulting taxonomy of sounds is tripartite: *Geophony* represents earth-related sounds, such as those of winds, thunders, or earthquakes. *Biophony* refers to sounds that are produced by nonhuman organisms that populate a biome. Krause remarks that biophony is rooted in geophony because the sounds in this category not only seem to be compartmentalized across species that share a habitat but can also bear direct similarities with the sounds of the earth, evincing mimicry. Finally, *anthrophony* represents the sounds produced by humans. According to Krause, humans might have been inspired by geophonic and biophonic sounds when they first began to establish a catalog of vocalizations: "Through mimicry, we would have transformed the rhythms of sound and motion in the natural world into music and dance—our songs emulating the piping, percussion, trumpeting, polyphony, and complex rhythmic output of the animals in the places we lived" (2012: 127).

Krause's theory is supported by a rare ability of humans among mammals: vocal reproduction of sounds heard by the ear (Merker 2012). This ability, called vocal learning, is exclusive to humans among primates and observed in only few other species, including songbirds and parrots (Patel 2007: 361). Although the field of biomusicology is ambivalent on the connection between music and animal song (Brown, Merker, and Wallin 2001: 8), the prerequisite function of vocal learning in both song and speech (Fitch 2006: 175) illuminates another thread in the evolution of musical behavior. Many theories in biomusicology suggest that music might have coevolved with or even originated from speech and language (Marler 2001: 45; Richman 2001: 301). But despite the similarities between music and speech in terms of how they communicate emotion, the evolutionary origins of the association between pitch patterns and emotional states are not clear. These associations may have emerged arbitrarily and evolved over time into a formal communicative code, or they might have formed based on the physiological traits of the human auditory system (Curtis and Bharucha

2010: 346). Regardless of the origins of such similarities, a mode of expression shared between music and language is gesturality, owing to the fact that both speech and music are tightly intertwined with movement (Mithen 2005). This view reinforces the common understanding that music is not merely an auditory experience. Indeed, the auditory system functions in a multisensory context as evidenced by numerous studies that highlight the correlations between auditory, visual, haptic, and kinesthetic senses (Warren, Kim, and Husney 1987; Bradley 2000; Merer et al. 2007; Özcan and van Egmond 2009; Vines et al. 2011).

In an experimental analysis of how low-level variables in music, such as pitch, dynamics, and rhythm, are mapped to nonauditory representations of space and motion, the musicologists Zohar Eitan and Roni Granot draw conclusions that corroborate certain widely shared assumptions about the relationship between music and movement, such as the correspondence between an increase in pitch and upward motion. However, they also find that these types of cognitive mappings between sensory modalities are multifaceted and much more complex than previously assumed. For instance, a listener's association of a musical gesture to kinetic movement turns out to be rarely symmetrical: a decrease in pitch moves not only downward but also "leftward and closer" (Eitan and Granot 2006: 242).

In another experimental study on the dynamic similarities between music and movement, Sievers et al. (2013) observe that structural features such as rate, direction, and dissonance evoke similar emotions when they are presented to subjects in a musical context versus when they are shown as features of a visual animation. Furthermore, the study shows consistent results across participants from the United States and Cambodia, implying cross-cultural roots to how movement in music is interpreted by listeners.

Such findings bring up another major focus in biomusicology research, that is, the study of musical behavior in humans from birth both within and across cultures. While musical traditions can exhibit marked differences from one society to another, several studies indicate that some aspects of musical behavior may have evolved similarly across cultures. And although empirical analyses do not reveal musical universals that can be deemed absolute, there are at least those that are statistically significant (Savage et al. 2015). Among such universals are the perception of octaves as equivalent pitch qualities, division of scales into seven or fewer pitches per octave, divisive patterns of twos and threes in rhythm, and association of loudness, acceleration, and high-registered sound patterns to emotional excitement (Brown, Merker, and Wallin 2001: 14).

The psychologist Sandra Trehub enumerates other musical universals such as the prevalent emphasis on global structure in music, preference of small-integer frequency ratios between pitches, and the ubiquity of unequal steps in scales in different musical cultures (2001: 427). The universality of these characteristics implies biological roots to music, yet "most of these features are low-level structural properties" (Sievers 2013: 70). Acknowledging the concern among ethnomusicologists about the implication of biological determinism in musical universals, Brown, Merker, and Wallin shift the discourse around this topic toward a middle ground between biology and culture by highlighting contemporary biomusicology research that promotes a balance between "genetic constraints on the one hand, and historical contingencies on the other" (2001: 13).

From Biology to Culture

Even a cursory analysis of music history would show us that music has evolved into a much more complex phenomenon than discrete variations in low-level acoustic properties such as pitch and rhythm. Furthermore, some of the seemingly instinctual behaviors that are considered to be genetically transferred may be rooted in a naturally selected ability to learn. This theory, called *the Baldwin effect*, suggests a cultural inheritance of learned behavior across generations (Depew 2003: 3). More importantly, although biology does inform our musical abilities, social factors play a greater role in the development of musical behavior (Blacking 1973: 46).

Beyond the qualities of music that we appreciate on account of our physiological disposition as a species, there lies an experiential complexity that is rooted in culture. A multidisciplinary investigation of this complexity becomes necessary to form a comprehensive understanding of what makes music appealing. For instance, in a study conducted with infants, Trainor, Tsang, and Cheung observe a preference for consonant intervals between pitches (2002: 187). While this implies a biological origin for musical consonance, it also reveals that the aesthetic appeal of the so-called dissonant intervals is a cultural phenomenon (Curtis and Bharucha 2010: 346). Several other studies conducted with young children highlight a corroborating view that musical expectations heavily depend on cultural learning and do not develop fully until sometime between the ages five and eleven (Juslin and Västfjäll 2008: 569). In other words, we *learn* how to appreciate many qualities of music. Recent neuroscientific research supports this hypothesis. In an fMRI study conducted to understand the neural processes related to reward in previously unheard music, Salimpoor et al. observed significant effects of sociocultural factors related to experience and memory:

> Our appreciation of new music is likely related to (i) highly individualized accumulation of auditory cortical stores based on previous listening experiences, (ii) the corresponding temporal expectations that stem from implicit understanding of the rules of music structure and probabilities of the occurrence of temporal and tonal events, and (iii) the positive prediction errors that result from these expectations. (Salimpoor et al. 2013: 218)

These findings suggest that our musical tastes are based on the accumulation of our listening experiences. Whatever music we have heard thus far will contribute to our appreciation of the music that we will listen to in the future. In her 1972 book *An Individual Note of Music, Sound and Electronics*, Daphne Oram emphasizes the significant role of memory in our appreciation of music when she describes how our memories allow us to establish a deep and personal regard for certain works that we return to with increasing pleasure (78). These perspectives can also help explain the aversion of the "stubbornly conditioned ears," as Varèse refers to them, to new sounds in music. The musical conditioning of our ears evolves through our encounters with new music in ways that reinforce or challenge what we already appreciate in music. In his seminal lecture "Four Criteria of Electronic Music," Stockhausen (1972) highlights this transformative nature of sounds: "Whenever we hear sounds, we are changed; we are no longer the same after having heard certain sounds. And this is more the case when we hear organized sounds, organized by another human being: music."

The Material and Language of Music

Unlike everyday sounds, which are the natural byproducts of events in the environment, musical sound in the traditional sense is fabricated. The musicologist Vladimír Karbusický characterizes musical sound as a stylization that came into being as "a magnificent shaping" of acoustic material that has been developed over the course of centuries, and one which does not exist in nature outside of a few instances (1969: 41). Musical instruments are crafted to produce certain sound qualities that serve specific musical functions.

Here, we are reminded of Krause's (2012) *The Great Animal Orchestra*, where he describes how biophonic sounds (i.e., nonhuman biological sounds) are compartmentalized in nature. In his book, Krause recounts having discovered a "biophonic score" while analyzing a soundscape recording, wherein the sounds of different animals occupied adjacent portions of the frequency spectrum. This natural allotment of frequencies enables individual species—such as bats, insects, hyenas, and elephants—to communicate within dedicated bandwidths. Indeed, much like the animals of a supposed biophonic orchestra, musical instruments take up different parts of the frequency spectrum based on their size, shape, and material. The violin plays in a higher register than the cello, which itself sits above the double bass on the spectrum. The instruments of the classical orchestra are designed to complement one another with unique sonic and expressive affordances. The acoustic properties of an instrument inform the musical range that it is best suited for, while the ergonomic limitations of the instrument dictate the kinds of musical phrases that can be performed with it.

Most acoustic instruments are designed to produce harmonic sounds with smooth temporal variations, whereas the sounds we hear in our daily lives are much more complex both spectrally and dynamically (Gaver 1993a: 2). Although the extended techniques that we encounter in experimental performance practices can uncover new musical affordances of traditional instruments, most of these instruments are originally conceived to articulate proportional tonal relationships. That being said, the musical traditions and the instruments of a culture continually influence and reinforce one another: an instrument can be designed to fulfill the needs of a musical tradition, but as the instrument gets refined over time, it begins to shape the very musical tradition it sprouted from.

The history of the Turkish bağlama reveals one such relationship. With the introduction of metal strings around the thirteenth century, the kopuz, a horsehair lute traditional to the Anatolian region, began to go through an evolution. The use of metal strings in this lute enabled the attachment of additional strings to the instrument, which in turn necessitated the use of a plectrum (Parlak 1998). The increased tension introduced by the strings and the added physical force of playing with a plectrum prompted a transition from leather to wood in the body of the instrument. This switch facilitated a new technique, where the performer could tap the body of the instrument to create percussive sounds. Combined with these percussive affordances of the wooden body, the use of metal strings with new timbral and dynamic properties ushered in a whole new musical tradition in Turkish folk music.

In a similar example, the evolution of the piano from the harpsichord into the fortepiano and eventually into its current form has brought about a marked increase in the dynamic range of the instrument, which in turn contributed to the wide dynamic range that characterizes the Romantic-era compositions (Lin 2012). Such examples demonstrate an inherent reciprocity between the materials and languages of music.

The concept of a musical language in itself is a cause of contention. Whether music has (or is) a language is a long-standing debate that is, to some extent, fueled by the divergent views on what constitutes a language. Music is commonly framed, much like any language, as "a mode of interacting with others" (Cross 2010: 7). The anthropologist Claude Lévi-Strauss has famously described music as "the only language with the contradictory attributes of being at once intelligible and untranslatable" (1975: 18). Indeed, translatability is often referred to as a quality that distinguishes language from music (Jackendoff 2009: 197). On the other hand, researchers have identified common mental mechanisms underlying language and music to evince an overlap between the two. The fact that the human brain maps musical sounds to internal structures has been characterized as a language-like property of music (Sloboda 2005: 177). For the psychologist Aniruddh Patel, music and language are both human universals that are based on the organization of discrete elements (e.g., melodies and words) into hierarchically structured sequences (2003: 674). While the two may not operate at the same level of syntactic representation, the ways in which the linguistic and musical syntaxes are processed are more similar than we might think (676). In a neuroscientific study on the overlap between music and language, Koelsch et al. used a neural signature that represents the compatibility between a word and its context as a measure to demonstrate that music can evoke semantic concepts much like language. Using music–word pairings gathered from a previous study, the researchers presented listeners with related and unrelated pairs of musical excerpts and words. Looking at an electrophysiological index of semantic priming, they observed an effect similar to what occurs between words and sentences, discovering that music can communicate "considerably more semantic information than previously believed" (Koelsch et al. 2004: 307).

Although instrumental music does not convey explicit meaning to the same extent as a conventional language, the interaction between a performer and a listener can be viewed as a form of communication, whether this involves extramusical associations, abstract thoughts, or feelings. Adopting a broad interpretation of language as a system of communication based on shared conventions, it could be argued that music employs a language rooted in the "codes of a convention consolidated over the centuries" (Boulez 1986: 4). As soon as we deem a series of sounds as music, we draw on a different set of conventions than we use for interpreting everyday sounds; we evaluate the musicality of a sound by the extent to which it complies with the musical precedents that we are familiar with (Meelberg 2006: 15). Each musical tradition has its own conventions that establish a language for the instruments of that tradition to *speak*.[1] This language is culturally fabricated much like the instruments themselves. The notational systems that represent these languages grow out of, and subsequently reinforce, a widely shared consensus on what counts as music (Prieto 2020).

Musical conventions, such as melodic, harmonic, and rhythmic structures, are abstract constructs that have been gradually established over time. As the aforementioned biomusicology research demonstrates, some of these conventions, including metric and intervallic proportions that are favored universally, are rooted in human biology. For instance, frisson (i.e., the feeling of getting chills) is a musical affect that we experience in response to loud passages. The musicologist David Huron argues that frisson is an innate response to surprise: a loud sound is symptomatic of high energy, which in turn is indicative of biological danger (2006: 35). A loud musical passage can therefore make us feel the thrill of danger without being in a dangerous situation. But beyond such biologically driven musical conventions, we have constructed a plethora of clichés to facilitate musical communication (2). In doing so, we fabricated musical structures that have been engraved in our deep-seated mechanisms of music perception. It is these kinds of musical conventions that, in some sense, reverse the traditional development of a semiotic relationship, wherein a referent (e.g., a word) is created to point to an external object or concept. In musical conventions that develop culturally, the sign (i.e., the abstract constructs of a musical language) come to synthesize the referent (i.e., affective appraisal). The sociocultural environments we are born into provide us with an initial context of conventions, and as we become exposed to new music, our musical vocabularies expand.

---

[1] Adorno and Gillespie describe playing music as speaking its language (1993: 403).

Music and Emotion

The combined sensory and cognitive experience of music influences the listener's affective state (Salimpoor et al. 2013: 218). Although music does not hold an immediate survival value, it activates brain mechanisms that process pleasure and reward, particularly those that are involved in the formation of learned associations and representation of value in stimuli (Omar et al. 2011: 1814). Through complex interactions between auditory, attentional, motor, emotional, and cognitive mechanisms, a transition from listening to feeling takes place (Chapin et al. 2010: 11, 13).

In their analysis of emotional responses to music, the psychologists Patrik Juslin and Daniel Västfjäll argue that the mechanisms through which music evokes emotion are not unique to music (2008: 559). Emphasizing a need to investigate the processes that underlie the emotional appraisal of music, they enumerate six psychological mechanisms that they consider to be involved in this appraisal: brain stem reflex, visual imagery, episodic memory, evaluative conditioning, musical expectancy, and emotional contagion. These mechanisms do not function in a mutually exclusive manner but rather assume complementary roles when processing emotions; our emotional experience of music is the result of the interactions between these mechanisms (572).

The *brain stem reflex* deals with the low-level structural and cross-cultural properties of the musical experience. These hard-wired reflexes are tied to the early stages of auditory processing. Sounds that are sudden, loud, or dissonant and those that feature fast temporal patterns signal the brain stem about potentially important events and induce arousal. This type of arousal reflects the impact of auditory sensations where music functions as sound at a primitive level (564).

Visual imagery is a mental process that is akin to perceptual experience but occurs in the absence of relevant stimuli to trigger perception. Although several studies have suggested that music can evoke mental imagery (Taruffi and Küssner 2019: 62), whether the associated mental processes are pictorial or propositional is an ongoing ontological debate (Juslin and Västfjäll 2008: 566). However, listeners do seem to conceptualize musical structure through metaphorical mappings between music and image schemas that are grounded in bodily experience (ibid.).

Music that accompanies certain life events are stored in *episodic memory*, which can be pivotal in discerning whether a musical experience is pleasant or not. So, even though episodic memory is subjective, it is still an important source of emotion when listening to music (567). Whereas episodic memory forms a cerebral connection between life experiences and musical emotion, *evaluative conditioning* is a process in which music is associated with an emotion as a result of its repeated co-occurrence with other stimuli of negative or positive valence.

Musical expectancy pertains to the induction of emotions through the violation of a listener's expectations. This, however, should not be confused with simple forms of unexpectedness, such as the sudden onset of a loud tone, which would instead trigger a brain stem reflex. The development of musical expectancy involves the cultural learning of syntactical relationships across the components of a musical structure (568).

Finally, *emotional contagion* occurs when a listener perceives an emotional expression in music and mimics this expression internally, leading to an induction of the said emotion. Research has shown, for instance, that music with slow tempo, low pitch, and low volume can induce an expression of sadness through emotional contagion (Juslin 2001).

In their open peer commentary on Juslin and Västfjäll's proposal, Fritz and Koelsch suggest the addition of *semantic association* to the list of mechanisms that underlie music perception (Juslin and Västfjäll 2008: 580). They give an example from African music where the predefined meaning of certain drum figures can evoke semantic associations. Since the semantic concepts attached to such figures exhibit emotional connotations, the decoding of these associations can induce emotional responses (580).

The musicologist Leonard B. Meyer refers to musical structures which elicit semantic associations as *connotative symbols*. A connotative symbol can designate a specific emotional state and be indicative of "a specific idea, concept, or individual" (Meyer 1956: 260). Meyer's comprehensive analysis of meaning in music constructs—and subsequently deconstructs—a dichotomy between absolute and referential meanings in music. The absolutist approach claims that music communicates only musical meaning, which is *abstract* and *intellectual*, and that musical material refers to nothing but itself. Conversely, the referentialists contend that music can communicate meanings that indicate extramusical concepts, actions, emotions, or characters (1). While Meyer agrees that music can indeed convey referential meanings, he remarks that the absolutist and referentialist camps are not mutually exclusive: These two types of meaning "can and do coexist in one and the same piece of music, just as they do in a poem or a painting." More importantly, both types of meaning depend upon learning (2). Meyer suggests that the communicated meaning of either flavor can be intellectual and emotional at the same time (39), corroborating some of the aforementioned studies on appraisal of music.

Affect in Music

As we discussed earlier, the translation of musical material into emotion is a multifaceted process that utilizes a host of interconnected perceptual mechanisms. One of the earlier stages of this process is the experience of affect. Affect has been applied to studies of human experience across a variety of domains including virtual reality (Bertelsen and Murphie 2010), painting (Deleuze 2003), politics (Massumi 2010), and sports (Ekkekakis 2012). This concept has not only been adopted in a large array of disciplines but also been the subject of diverse interpretations. While some use affect and emotion interchangeably (Shouse 2005; Lim et al. 2008: 118), others utilize it to characterize those aspects of musical experience that extend beyond basic human emotions (Jackendoff and Lerdahl 2006: 60).

In philosophy, the use of affect dates back to Spinoza's *Ethics*, where he described it as an affection of the body by which "the body's power of acting is increased or diminished" (1994: 154). As a physical disturbance of the body, affect precedes subjective feelings (Wetherell 2012: 3). In his introduction to Deleuze and Guattari's *A Thousand Plateaus*, the philosopher Brian Massumi characterizes affect as a pre-personal intensity that evokes a transition of the body from one experiential state to another (in Deleuze and Guattari 1987: xvi). The music philosopher Vincent Meelberg adopts Massumi's interpretation of affect to articulate a difference between musical gestures and sonic strokes. Meelberg describes a sonic stroke as an acoustic phenomenon that induces musical affect upon impacting the listener's body (2009: 325). A consequence of this impact is emotion, which emerges once the affect is reflected upon, that is, once a sonic stroke is registered as a musical gesture.

While philosophical inquiry may not always map onto empirical research, the perceptual traits of affect show similarities with those of the brain stem reflex. A functional coherence between affect and the brain stem reflex is highlighted in their intrinsic reliance on the perceptual properties of musical sound. Due to its role in the early stages of auditory processing, the brain stem reflex is tightly coupled with the physiology of the perceiver and the so-called universals (i.e., the low-level structural properties) of musical experience. While affect represents the corporeal segment of the affective appraisal of music, much like the brain stem reflex, it is tightly coupled with the ensuing interplay between the mechanisms that underlie music cognition. Emotion, on the other hand, is a personal *capture* of affect—an intensity that is qualified and inserted "into semantically and semiotically formed progressions, into narrativizable action-reaction circuits, into function and meaning" (Massumi 2002: 35). This continuity between affect and semantic processing, which is also reflected in Meelberg's discussion of the relationship between sonic strokes and musical gestures, supports the understanding that the so-called subjective world of mental representations stems from our embodied interactions with our physical environments (Leman 2008: 13). Sharing many mechanisms with perception and action, our conceptual abilities are grounded in sensorimotor simulation (Pecher, Zeelenberg, and Barsalou 2003: 123).

This view brings us back to the understanding of music cognition grounded in the indivisible union of the body and the mind—biology and culture. From a phenomenological perspective, mechanisms of anticipation are considered to mediate our everyday experiences (Schutz 1967: 58), musical or otherwise. This is why the cognitive musicologist David Huron characterizes musical expectancy as being interwoven with both biological adaptation and cultural norms (2006: 357). In Chapter 4, I will give an overview of studies that underscore the cooperation between the perceptual and cognitive faculties in forming the mental representations through which we parse our everyday experiences. Then in Chapter 5, we will revisit the concept of affect as it applies to our embodied engagement with electronic music, exploring the affective qualities of the mental representations evoked by electronic music. But before all of that, let's focus our discussion of musical appraisal on our engagement with electronic music.

Experiential Idiosyncrasies of Electronic Music

In this section, we will explore how electronic music diverges from instrumental music in terms of the listening experience. There are two standpoints that are fundamental to this discussion: Firstly, although I initially use a dichotomy between these two types of music as a device to highlight the experiential idiosyncrasies of electronic music, we will ultimately discover a continuity between the two as the material brought to bear by the electronic medium expands the vocabulary of instrumental traditions. Secondly, when I refer to the listening experience, I do not draw a distinction between that of the composer and that of the audience. As I outline the cognitive qualities of the electronic music experience, I will situate listening as the operative act in not only the appreciation but also the creation of electronic music. In the previous chapter, I mentioned that the electronic medium affords a host of techniques for sound design and music composition. I then remarked how two pieces created using two completely separate techniques can be perceived by the listener as being similar in many ways. This also implies that the listener can infer an entirely different compositional plan than what the composer has conceived, if such a plan had existed in the first place. To resolve such conflicts and overcome the supposed communicational hierarchy between the composer and the listener, I will adopt a semiotic model that places the emphasis on listening as an intrinsically creative act.

The Composer, Who Is Also a Listener

The composer Horacio Vaggione characterizes the idea of "the composer as a producer" as a direct correlate of "the composer as a listener" (2001: 60). The composer's act of listening can take different forms over the course of the creative process. In this context, Vaggione identifies two main modes of hearing: An *inner hearing* takes place during the silent writing phase, which constitutes a sizeable portion of instrumental music composition; the composer fleshes out musical ideas in the form of mental representations that are yet to be materialized. When the composition process eventually begins to yield audible results, a *concrete hearing* occurs. For the composer working in the electronic medium, the inner and the concrete modes of hearing operate simultaneously; while the interplay between these actions are unique to each instance of composition, they nevertheless coexist. The dichotomy between these two modes is reminiscent of the philosopher Don Ihde's distinction between the inner and outer experiences of auditory phenomena (2007: 119). Ihde suggests that the inner experience (i.e., the imagination) can be refined and enriched as it gets contrasted with the outer experience (i.e., the perception). As I will detail shortly, measuring the machinations of inner hearing against auditory artifacts becomes a reciprocal and immediate process in electronic music composition.

The gamut of compositional strategies in electronic music is vast. For the sake of discussion, let me portray a specific scenario wherein the creative process begins with inner hearing: The composer *imagines* or *recalls* a sound. This sound may have been conjured up as part of a premeditated structure, such as a theme or narrative that incorporates this sound, or it might serve as a starting point at a more exploratory stage of composition where the sound itself inspires a compositional plan. The composer might have heard this sound before or they might be making it up from scratch, but the field of possibilities is, on the one hand, demarcated by the limits of their auditory perception and, on the other, seeded by all the sounds they have previously heard. The composer might choose to notate this sound; this can take the form of traditional music notation or just an inscription on paper. This inscription might represent the temporal unfolding of an envelope in the style of a waveform plot or portray the spatial movement of a sound object. Such drawings can display similarities with how we talk about sounds using hand gestures, for instance, when we raise our hand to indicate an increase in pitch or move it from side to side to illustrate panning. Two examples of this kind of notation are seen in Figure 2.1.

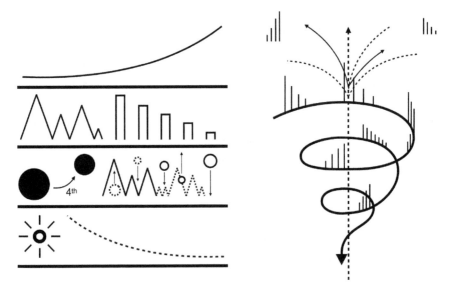

Figure 2.1 Digitized excerpts from my notations for two pieces. On the left is a notation of the opening segment of *Diegese* until the 0′16″ mark. The visual markings indicate multiple sonic properties including time, pitch, amplitude envelope, and timbre. On the right is a notation of the segment between 0′34″ and 0′40″ in *Christmas 2013*. While the vertical and horizontal axes are mapped to pitch and panning, respectively, the spiral represents the three-dimensional movement of staccato impulses, with the arrowheads indicating the temporal progression of gestures.

The composition process so far has been silent. A sound has been imagined and it may have been represented in writing, but these actions have been solely guided by the inner hearing. This is not intrinsically different from instrumental music composition, where creating visual representations of musical ideas corresponds to the writing of the score. This process, however, may remain the instrumental composer's sole activity until the work is interpreted by a performer. Until then, the back-and-forth between creating and evaluating the work transpires within a loop between the composer's inner hearing and the score on paper. The composer can audition a section of the work by playing it with an instrument or a software application, but this will be a mere representation of the actual work. The said work could also be a piece for a solo instrument that the composer is capable of performing by themselves. In this case, either the inner hearing lends itself to concrete hearing in the form of an improvisation wherein the writing is no longer silent, or the composer assumes consecutive identities, writing music through inner hearing as a composer and later interpreting as a performer.

Electronic music blurs this distinction by affording a coexistence of hearing modes. To illustrate this, let me continue describing the compositional scenario from where I left off: A sound has been imagined through inner hearing, though it has not been materialized into an acoustic phenomenon. At this point, the composer might begin to design the sound, making the very act of composition audible. There are numerous strategies for executing this: If the sound in mind is referential (e.g., the sound of a wind blowing, the sound of a glass shattering), the composer might record this sound from an actual source or record other sounds that create an illusion of the reference in style of foley in filmmaking. They can also generate sounds that create an illusion of the reference. In the latter method, the composer might investigate the physical characteristics of the sound in mind: What are the temporal and spectral properties of this sound? How does it resonate or move within space? Which synthesis methods would work best to recreate the sonic qualities of this sound? If there are existing recordings of this sound, the composer might analyze the auditory invariants in these recordings that make the source of the sound recognizable. Based on this analysis, the composer might use a combination of oscillators, noise generators, amplitude envelopes, and filters to *sculpt* the sound, or if the sound is the product of a complex process, they might build a piece of software that can emulate the unfolding of this sound.

On the other hand, if the sound in mind is purely speculative without a clear reference, the composer might first evaluate the complexity of the imagined sound. Is it easier to generate this sound, or could it be micro-montaged into existence using bits and pieces of recordings? Even if the sound itself is not referential, what could be a reference *to* this sound? What is the sound of that reference, and how does it relate to the composer's speculation? Then the strategies listed above come into play: the composer could record or synthesize sounds, either manually or via algorithms, in order to materialize the sound that they have conceptualized.

In electronic music, these actions prompted by the composer's inner hearing will have immediate sonic outcomes that will instigate concrete hearing. The outcome of these actions will not be a mere representation of the musical material but will make up the actual work itself without the need for a performance that mediates between the inner and the concrete acts of hearing. The electronic medium provides composers, who have traditionally had to seek help from an intermediary language (i.e., the notation) and an interpreter (i.e., the performer), with immediate control over sounds (Oram 1972: 91). The composer Maryann Amacher describes that this kind of control enabled her to experientially discover the compelling sonic perceptual features that she employed in her electronic works (2008: 10). Through such control, the composer can conceive a sound and experience it at the same time. For Morton Subotnick, this idiosyncrasy of electronic music was a primary motivation for him to begin composing for the electronic medium in the 1960s:

> I thought what I saw was the possibility one day of having a studio in your home and to create a whole new music, where you would be where music could become a sound. A studio art, where I could have an idea and try it out. Instead of putting it on paper and having musicians play it, I could actually try it out directly, listen to it and redo it, just like a painter would. ... I would be the composer and the performer and the listener, then send it off, and other people could listen to it. (Rosenbloom 2011)

Subotnick's painting analogy is particularly effective in highlighting the concurrency of the ideation and the materialization of sounds in electronic music. Vaggione characterizes this immediacy of the interplay between the composer's actions and their audible outcomes as an "action/perception feedback loop" (2001: 60). To gain a better understanding of this concept, let's imagine a scenario in which a composer designs a sound that does not have an external referent. Upon the first step that the composer takes to produce this sound, their inner hearing is instantaneously supplemented by their concrete hearing. As they proceed to develop this sound, their initial conceptualization of the sound begins to evolve. This is the result of a complex negotiation, which will be unique to each sound. The sound imagined via inner hearing can take total precedence and steer the design process. Conversely, the sound which the composer brings into physicality can begin to overpower the imagined sound by offering new possibilities previously unimagined; the composer, who is also a listener, discovers for the first time the embodied qualities of their quiet speculation. Every action will trigger new perceptions, which, in return, will influence the composer's next step. Individual sounds, and eventually the composition as a whole, will emerge through this immediate interplay between actions and perceptions:

> In a manner similar to a child's cognitive development, active manipulation of the "object" leads to its functional and conceptual understanding, and therefore, similar to its role in childhood, play is an important activity to stimulate that learning. Playing with a sound involves both memory and imagination, the "what if" question, and the sense of discovery. (Truax 1996: 60)

A distinct contribution of the modern electronic medium to this process of play is its affordance of extraordinary degrees of accuracy and elasticity. For instance, the level of temporal precision that modern computers offer is virtually unlimited within the bounds of human perception. We can zoom into an audio file and edit lengths that are a fraction of a millisecond. These, and many other digital audio editing techniques, such as stretching, duplicating, and reversing, are relatively recent implementations that we now view as standard features of audio software. Such processes have become inextricable from modern composition workflows, bringing an unprecedented level of elasticity to the sonic material.

This level of elasticity of sound material was not at the artist's disposal up until the late 1980s when the first digital audio workstations were introduced. Even then, hardware constraints significantly hindered the capabilities of audio software. As someone born into the era of personal computers, even I am able to look back and identify remarkable leaps in computing power that had drastic effects on my work as a composer. These leaps have especially become apparent in the quality and efficiency of resource-demanding processes such as spatial and frequency-domain transformations. The number of such processes that can be applied to an audio track and the concurrent number of effect-heavy audio tracks that can be played back in real time without disruptive CPU overloads have increased substantially. Modern personal computers provide composers, sounds designers, and audio programmers with an abundance of hardware capacity. The implications of such processing power go beyond a feat of hardware specifications. The technical gap between a composer's concept and its perceivable outcome is continuously shrinking: the ever-expanding arsenal of audio processing tools makes it progressively easier for composers to materialize sounds that are faithful to the fantasies of their inner hearing. This is why Roads refers to the twenty-first century as "the golden age of electronic music" (Robindoré 2005: 11).

From Parameters to Instincts

A historical overview of the interplay between technology and the composer shows us how certain characteristics of electronic music have emerged over time. Although there had been accurate predictions of the electronic medium's impact on our musical vocabulary, the material it unlocked still caused a musical paradigm shift. The composer Pierre Boulez recounts how guided explorations in the early days of electronic music brought about a wide range of unsuspected possibilities, so much so that new mental categories had to be created before composers could make meaningful use of this material (1986: 9).

As we discussed in Chapter 1, the early theoretical perspectives on electronic music guided composers through their initial encounters with the electronic medium in the 1950s. One of the most significant examples of this was the adoption of serialism from the Second Viennese School as an artistic direction for the electronic music studio at the Westdeutscher Rundfunk (WDR) in Cologne. One of the founders of this studio, Herbert Eimert, claimed that it would not have been possible to exert any level of musical control over electronic sound had they not embraced the vision of Anton Webern, who was the most influential exponent of serialism in the domain of instrumental music (Chadabe 1997: 37). The composers of the Cologne studio exercised a particular flavor of this technique called total serialism, where serial permutations are applied to not just pitch sequences but also other parameters, such as duration and dynamics. There was indeed a natural marriage between total serialism and the electronic medium. On the one side was a composition technique that demanded meticulous control over every parameter of musical sound including pitch, rhythm, loudness, and timbre; an ideal materialization of the finely pointillistic essence of this style required superhuman performance. On the other side was a brand-new compositional medium that afforded an unprecedented level of mathematical precision over the control of sonic parameters via the use of electronics.

Most composers, however, did not have preconceived ideas as to how devices such as oscillators and signal processors could be utilized within the context of music. The studio technician played a significant role in bridging the gap between the artist and the medium. As a result, most composers were not in total control of the creative process. Furthermore, technical challenges were coupled with a lack of aesthetic precedent. An "irrational necessity" to overcome the stagnation in the world of musical instruments came before aesthetic reflection, which was all but relinquished in favor of free development (Boulez 1986: 9). The inherent compatibility of the electronic medium with the parametric nature of serialism, and the existing affiliation of the Cologne composers with this style in the instrumental domain, represented intuitive segues into electronic music. When Eimert's earlier statement is viewed in this context, it could be argued that serialism functioned as an initial comfort zone for the composers facing the unknowns of the electronic medium for the first time.

In the previous chapter, we discussed how the strictly deterministic reliance on serial permutations exercised during the early years of the Cologne studio was criticized by the members of the Paris studio. Conversely, the stylistic direction of the Paris studio, which relied heavily on exploration and play, was disparaged by the adamant practitioners of serialism for lacking any deterministic basis. In 1955, a few years after the studios in Paris and Cologne were founded, the composers Luciano Berio and Bruno Maderna established the Studio di Fonologia Musicale in Milan. They instated a mix between serialism and musique concrète as the studio's stylistic orientation, referring to it as *radiophonic art*, which relied on the said styles only to the extents deemed necessary by the artist.

According to the composer Luigi Nono, who succeeded Berio as the head of the studio, any artistic dependence on mathematical relationships without interrogative contemplation could deprive art of its reason to exist (1960: 1). On the other hand, he also criticized the delegation of artistic determinism to randomness, characterizing it as a sign of an inability on the composer's part to make decisions. Nono's arguments emphasize the significance of creative initiative and the conscious translation of artistic instincts to actions. For Nono, neither pure determinism nor pure randomness was supposed to overpower intuition and logic (3). It is worth noting that deterministic procedures governed a significant portion of Nono's early work; the composer merely opposed lending too much authority to either approach. Pierre Boulez, who was a prominent practitioner of serialism in the early 1950s, similarly emphasized a need for a middle ground:

> Musical invention must bring about the creation of the musical material it needs; by its efforts, it will provide the necessary impulse for technology to respond functionally to its desires and imagination. This process will need to be flexible enough to avoid the extreme rigidity and impoverishment of an excessive determinism and to encompass the accidental or unforeseen, which it must be ready later to integrate into a larger and richer conception. (Boulez 1986: 11)

As Boulez further points out, when the materials and the ideas underlying an artistic work develop independently, an imbalance growing between the two can have an adverse effect on the work (6). This is mitigated by an ability to think through the creative possibilities of the electronic medium—an ability that is inherently tied to the artist's grasp of technology. As early composers of electronic music developed this ability, their creative instincts began to converge with the affordances of electronic sound.

Karlheinz Stockhausen, who produced arguably the most prominent works of electronic music to come out of the Cologne studio, succeeded Eimert as the artistic director of the studio in 1963 and remained at the helm until 1977. Stockhausen's oeuvre of electronic music constitutes an illuminating timeline of both the stylistic evolutions in early electronic music and artists' internalization of technology. His first experiments at the WDR involved serialist studies that closely followed the stylistic direction of the studio. Between 1955 and 1956, within five years of the studio's foundation, Stockhausen composed *Gesang der Jünglinge*, which marked a departure from a strict adherence to serialism (Holmes 2008: 66). Then in 1959, he completed his electronic epic *Kontakte*, which relied on serial proportions only at a "broad formal level" (Toop 1981: 189). To realize this piece, Stockhausen devised new methods to create extended tape loops and quadraphonic sound, which he then employed to exploit the thresholds of human hearing. While the composition of this work depended on serial methods to a certain degree, Stockhausen had, by that time, gained enough artistic control over the electronic medium to override these methods when he deemed it necessary. In a 1997 interview, Stockhausen recounts an instance where he ended up recomposing an entire section of *Kontakte* after coming to the startling realization that the preplanned construction of this section did not yield an organic result when spliced together (Paul 1997). As the musicologist

John Dack (1999) describes in his article "Karlheinz Stockhausen's Kontakte and Narrativity," the composer did not allow "any rationalistic method to take precedence over musical instinct."

In some cases, the composer's divergence from a style took place at a more institutional level. Recounting her decision to leave the Paris studio in the late 1960s, Éliane Radigue describes how her music was viewed as "an injury toward the basic principle of musique concrète" (Rodgers 2010: 59). In order to pursue her own interests in electronic music, Radigue felt it necessary to separate herself from the studio. A few years later, she decided to move to New York to gain access to analog synthesizers, which were not available in Paris at the time (55). This decision paved the way for her pioneering work with the ARP 2500 synthesizer. Radigue describes her relationship with the electronic medium as follows:

> Then came the electronic Fairy; through the power of magnetic, analog and digital capture, breath, pulsations, beating, and murmurs can now be defined directly in their own spectrum, and thus reveal another dimension of sound—within sound. ... The freedom of a development beyond temporality in which the instant is limitless. Passing through a present lacking dimension, or past, or future, or eternity. Immersion into a space restrained, or limited by nothing. (Radigue 2009: 48–9)

Complexity of Listening

As composers firmed their grasp on the electronic medium, what was *heard* in electronic sound started to take precedence over what had been planned or *parametrized*. Halfway through the twentieth century, composers began to acknowledge that humans perceive electronic sounds as homogeneous phenomena rather than manipulations of individual acoustic properties (Stockhausen 1962: 40). During the ensuing decades, the mutual influence between parameters and the "complexity of listening" gained even further prominence in electronic music practices (Zampronha 2005). The musicologist Joanna Demers refers to this period as the *post-Schaefferian era*, where the creative interest in catering to reduced listening began to decline as external associations increasingly became an integral aspect of electronic music (2010: 14). The composer Ambrose Field expresses that composers and listeners no longer needed to disregard extramusical connotations in electronic music

(2000: 37). According to Demers, this evolution is, above all, practical, since both empirical evidence and common sense situate the recognition of sounds as an intrinsic function of listening (2010: 84).

The composer Trevor Wishart, while opposing the premise of reduced listening, agrees with Schaeffer on the possibility of maintaining control over the listening experience (Demers 2010: 30). However, he argues that a set of metaphoric primitives would be necessary to establish a complex metaphoric network between the composer and the listener. Wishart describes metaphoric primitives as "symbols which are reasonably unambiguous to a large number of people" (1986: 55). This definition reminds us of the concept of *connotations* described by Meyer as conscious image processes that are common to a whole group of individuals within a culture (1956: 257). Triggered by musical stimuli, image processes are memory functions that act as mediators through which music arouses affect. Connotations are based on the similarities between our experience of musical material and the "non-musical world of concepts, images, objects, qualities, and states of mind" (Meyer 1956: 260). The recognition of such connotations requires habituation and automatism, which are established over time through repeated encounters with a given association. It is worth noting that Meyer outlines this concept as it applies to instrumental music and acknowledges the difficulty of specifying connotations in that context.

On the other hand, the material of electronic music gives new significance to connotations and, more generally, semantic processes in music. Audio reproduction and synthesis techniques unlock for the artist an entirely new and virtually unlimited vocabulary of sounds. Unsurprisingly, the extent of this sonic material far exceeds the vocabulary of a traditional musical language and "pitches music into a no-man's-land" (Demers 2010: 13). As Smalley remarks, when music is opened up to all sounds, discovering a path through the "bewildering sonic array" of electronic music becomes a challenge for both the listener and the composer (1997: 107). The electronic music listener rarely has preconceived notions as to what to expect from a new piece as "everything remains to be revealed by the composer and discovered by the listener" (Smalley 1996: 101). Since anticipation is a decisive element of our conscious experience, musical expectancy plays a prominent role in how we interpret a particular piece of music. But if any sound within the thresholds of human hearing is to be expected in electronic music, how can the composer construct the unexpected?

Surprise requires an expected outcome; and an expected outcome requires an internalized norm. Composers must activate either normative schemas (such as styles) or commonplace clichés in their listeners if their violations of expectation are to have the desired effect. (Huron 2006: 36)

Supporting this view, Field argues that a common ground is necessary for the composer and the listener to communicate (2000: 40). This sentiment echoes Wishart's proposal of metaphoric primitives and Meyer's idea of musical connotations, both of which require a common framework for any communication to transpire between the artist and the audience; establishing a framework of anticipation for the experience of electronic music therefore requires a shared point of reference between these actors. A normative schema that pertains to our perception of sound is already encoded in our survival instincts. We are hardwired to perform what the composer Michel Chion refers to as "causal listening" (1994: 26), where one can identify a precise cause or, at the least, a category of sources for a sound. This reflex is rooted in the evolutionary disposition of our auditory system to locate and identify events; even in the absence of adequate information, "the perceptual system hunts" in its attempt to attach meaning to whatever information it can gather (Gibson 1966: 303). From this standpoint, all sounds can be considered to be signs: the perception of a sound indicates something beyond the sound itself as sound cannot exist as a pure abstraction (Demers 2010: 37).

This has various implications for the electronic music experience on both ends of the communication between the composer and the listener. The sound world of electronic music encourages the imagination of extrinsic connections (Smalley 1997: 110), making it "the first musical genre ever to place under the composer's control an acoustic palette as wide as that of the environment itself" (Emmerson 1986: 18). The material brought into the composer's vocabulary with the electronic medium sets the genre apart from its predecessors. This material motivates its own language—its own set of conventions for creating and parsing meaning—and amplifies the capacity of music to evoke external associations. As a result, the culturally and biologically encoded pathways from musical material to affective appraisal become susceptible to interception by a layer of meaning attribution rooted in the material of electronic music. A cognitive continuum ranging from abstract to representational becomes an instrument, which the electronic music composer can leverage to manipulate the listening experience.

Table 2.1 Language Grid (Emmerson 1986: 24)

Abstract syntax	1	4	7
Combination of abstract and abstracted syntax	2	5	8
Abstracted syntax	3	6	9
	I: Aural discourse dominant	II: Combination of aural and mimetic discourse	III: Mimetic discourse dominant

|———————————————— MUSICAL DISCOURSE ————————————————|

To illustrate the numerous ways in which composers can exploit this continuum, the composer Simon Emmerson formulates a language grid, as seen in Table 2.1. Between a mimetic discourse, which evokes images that are extrinsic to the musical material, and an aural discourse based on sound objects that are free of external associations, Emmerson portrays a continuous plane of possibilities (1986: 19). In pieces where a balance of both is found, although the sound sources are recognizable, "the impressions are welded together in other ways than those based on associative image" (20). The axes of the grid are discourse, which represents the degree to which mimetic references are used, and syntax, which can be either *abstracted* from the material or *abstract* in the sense that it is constructed independently from it. In devising this grid, Emmerson's goal is to address the divide between elektronische Musik and musique concrète, with the former relying on an abstract syntax (i.e., serialism) while the latter focuses on abstracting syntax from material (i.e., the plasticization of sound recordings into acousmatic phenomena). Each cell in the language grid corresponds to a combination of discourse and syntax, which Emmerson exemplifies with various pieces of electronic music. For instance, situated in Cell 1, Stockhausen's *Electronic Studies* from the early days of the Cologne studio pairs serialism as an abstract syntax with an aural discourse–dominant material (i.e., sounds synthesized with an oscillator). On the other hand, the composer Luc Ferrari's *Presque Rien*, situated in Cell 9, presents a mimetic discourse through the use of soundscape recordings, while drawing its syntax from these very recordings by allowing the mimetic discourse encoded in the material itself to unfold through the course of the piece.

An Amalgamation of Languages

A culturally established language of music in the traditional sense is often insufficient to codify the structural complexity of electronic sound. This, however, does not imply that electronic music is isolated from preceding musical traditions. As Karbusický expresses, the perception of electronic music is linked to those traditions through a logical evolution; according to the musicologist, electronic sound can embody all the possible functions of music in general (1969: 41). The materials and languages of instrumental music traditions are still inherent to many electronic music practices. Rhythmic and tonal relationships, and formal structures that have grown out of instrumental music traditions, often govern the design and organization of sounds in electronic music. In the composer Hanna Hartman's piece *Circling Blue*, soundscape recordings are intermixed with sounds of race cars and other streams of processed audio. As a foreground element, a singing voice intermittently emulates the sound of the race cars passing by as the pitch of their engines shift due to the Doppler effect. Here, the composer extracts a melody from a so-called nonmusical sound and contextualizes it as a musical line that accompanies soundscape elements that do not exhibit similarly traditional musical features. The composer Helmut Lachenmann's works involving the use of extended techniques on traditional instruments subverts the relationship between instrumental and electronic music practices in what he refers to as *instrumental musique concrète* (Heathcote 2003: 52).

As we will find out throughout the rest of this book, electronic music can instigate a vast range of experiences grounded in representational, abstract, conceptual, perceptual, semantic, and affective qualities of sound, among others. In the beginning of this section, I noted that although the current discourse might appear to impose a dichotomy between instrumental and electronic music, the ultimate point to be made would be that the language of electronic music does not supersede that of instrumental music, but rather expands it. This expansion may not seem intuitive on the outset; composers develop aesthetic strategies to internalize this expansion and formulate a continuity between musical traditions, old and new. This process, which I characterize as an amalgamation of musical languages, is best expressed by Edgard Varèse:

> My fight for the liberation of sound and for my right to make music with any sound and all sounds has sometimes been construed as a desire to disparage and

even to discard the great music of the past. But that is where my roots are. No matter how original, how different a composer may seem, he has only grafted a little bit of himself on the old plant. But this should be allowed to do without being accused of wanting to kill the plant. He only wants to produce a new flower. It does not matter if at first it seems to some people more like a cactus than a rose. (Varèse and Wen-chung 1966: 14)

Threads of Communication in Electronic Music

To delineate the implications of a so-called cognitive continuum on musical experiences, I will split the discussion into two threads of communication between the composer and the listener. But instead of simply contrasting the composer with the listener, I will adopt Jean Molino's semiotic model to describe these two threads through the acts of *poiesis* and *esthesis*. In this model, there are three dimensions to how symbolic meaning is communicated:

The poietic dimension is the sender's process of creation,

The esthesic dimension is the receiver's construction of a meaning from the poietic,

The trace is the material embodiment of the symbolic form, accessible to senses. (Nattiez 1990: 12)

This model is based on a restructuring of the classic schema of communication, which supposes a unidirectional transfer of meaning (i.e., producer → message → receiver), into a bidirectional model wherein the producer and the receiver act upon a neutral level (i.e., producer → trace ← receiver). This restructuring implies that a symbolic form (e.g., a poem, a film, a symphony) is not some intermediary step in a process of communication that transmits the meaning intended by the author to an audience. It is instead the result of a complex process of creation (i.e., the poietic process) that deals with the form and content of the work; it is also the point of departure for a complex process of reception (i.e., the esthesic process) that not only receives but also *reconstructs* a message (Nattiez 1990: 17). In other words, poiesis is the shaping of meaning (or concept) into material (the trace), whereas esthesis is the construction of meaning (or percept) from the trace. But *meaning as concept* and *meaning as percept*, as they apply to the same material, are not predestined to match.

Molino's ternary model casts an interesting light on the unique form of action/perception feedback loops inherent to electronic music composition. Poiesis and esthesis can easily be regarded as acts of the composer and the listener, respectively. However, since the electronic medium enables the coexistence of inner and concrete modes of hearing, composing via this medium can be formulated as a complex interplay between poiesis and esthesis. This perspective complements the aforementioned contextualization of *the composer as a listener* and frames the feedback loop between actions and perceptions as a dialogue. The momentary acts of composition can therefore be schematized as follows:

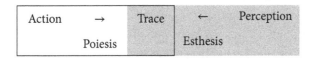

Following Molino's model, this schema conceptualizes the act of composition like a pendulum around the sonic artifact, putting equal emphases on the acts of producing and listening. The idea of a neutral level (i.e., the trace) not only promotes a democratized view of listening but also liberates the percept from the concept. This liberation can manifest itself during the composition process as illustrated in the above schema, or it can transpire between the composer and the listener. For the composer, a musical concept and its perceived outcome can contradict each other. This is an internal conflict between the acts of poiesis and esthesis. Conscious or unconscious resolutions of such conflicts guide the nonlinear progression of a composition process at various scales ranging from gesture to form. Communication of meaning between the composer and the listener is also fertile with such conflicts. In Chapter 4, we will further analyze these conflicts when we discuss the idea of gestural intentionality in electronic music. For now, let's examine the esthesic and the poietic threads individually in order to get a better sense of how listening, both that of the composer and that of the listener, acts upon the material that is the trace.

The Poietic Thread

In his *Aesthetic Theory*, Theodor Adorno describes art as the language of wanting the other: "The elements of this other are present in reality and they require only the most minute displacement into a new constellation to find their right position" (Adorno 2002: 132). An artwork is a demonstration of this displacement in reality. In her interpretation of Adorno's stance on artistic material, Demers deems it impossible to create an entirely new type of material from nothing; artistic material can either imitate empirical reality or mediate it by alienating the familiar through technology (2010: 50).

We have already discussed how our encounters with environmental phenomena establish a frame of reference for esthesic processes. This has self-evident implications for the poietic acts of the composer. A majority of a composer's knowledge of sonic phenomena is rooted in their experiences with their daily environments. When a composer designs a sound, the possible outcome is above all bounded by the physiological constraints of the auditory system. When the physical properties of an acoustic signal fall within the thresholds of auditory perception, the cognitive faculties come into play. The composer's ideas for designing a sound will be influenced by how the sounds they have encountered thus far have unfolded. This does not necessarily imply that every sound design is normalized to some form of representationality. But abstractness is nevertheless a function of concreteness; the unreal *becomes* through the real. In this vein, Adorno highlights the role of material in going beyond the material itself:

> The choice of the material, its use, and the limitations of that use, are an essential element of production. Even innovative expansion of the material into the unknown, going beyond the material's given condition, is to a large extent a function of the material and its critique, which is defined by the material itself. … If [the composer] turns critically against tradition through the use of an autonomous material, one completely purged of concepts such as consonance, dissonance, triad, and diatonicism, the negated is nevertheless retained in the negation. Such works speak by virtue of the taboos they radiate. (Adorno 2002: 148)

This perspective on the expansion of the material into the unknown through a negation of tradition is relevant to the relationship between abstractness and representationality in the arts. In an abstract work, physical reality becomes the negated; the process of negation relies on that which is negated during both the poietic and the esthesic acts. This principle applies to the representations of reality (or lack thereof) in electronic music as well. For instance, acousmatic music, in a similar fashion to Meyer's prioritization of embodied meaning in instrumental music, promotes a focus on the qualities of sound in itself rather than the external referents it may evoke. Even in the early years of the Paris studio, where this concept originates from, a common technique for producing acousmatic sound was the exhaustive investigation of sound recordings that were clearly indicative of their sources. By studying those attributes of a sound that facilitates its association with a source, composers discovered sound manipulation techniques for severing the material from such associations. These practices demonstrate that an extensive understanding of reality facilitates the abstraction of artistic material from that reality. In electronic music, the entirety of our auditory vocabulary serves as the reality to be embraced or negated.

But how does one meaningfully deal with such a large corpus? Exploration and experimentation are undoubtedly inextricable components of electronic music composition at any level of artistic expertise. However, as we further our involvement with the electronic medium, our strategies for materializing sonic ideas become increasingly innate. Through experimentation, we develop new techniques (and new clichés) that help us establish our unique voices as composers. Our inner hearing evolves as our auditory imagination begins to converge with our ability to hear through the possibilities of the electronic medium; we begin to think not just through sound but also through the medium.

When I was describing a composition scenario earlier in this chapter, I stressed the importance of the unimagined or unplanned in igniting new action/perception feedback loops. Exploring a sound material for creative leads is indeed a rewarding creative process that composers often turn to. That being said, the composer should not easily surrender to the immediate affordances and artifacts of the medium either:

> [Material] is everything that artists encounter about which they must make a decision. The idea, widespread among unreflective artists, of the open eligibility of any and all material is problematic in that it ignores the constraint inherent in technical procedures and the progress of material, which is imposed by various materials as well as by the necessity to employ specific materials. (Adorno 2002: 148)

The abundance of the creative opportunities available to the composer today can at times lure the composition process away from a kind of *guided exploration*. The temptation to exploit the "eligibility of any and all materials" can lead to a creative block disguised in unrestrained exploration. The composer might also give in to the constraints or the self-assertions of the medium by following the path of least resistance (e.g., by using presets in effects plug-ins or the default settings for time-domain and frequency-domain processes in a digital audio workstation [DAW]). The composer Alice Shields attributes this tendency to the ease and speed with which a composer can develop material using the electronic medium. According to Shields, when a composer caves in to the ready-made solutions that the medium affords, they run the risk of arriving at unremarkable timbres (Shields quoted in Lee 2018: 32).

In these situations, creative constraints can function as reflecting walls "inside which a tissue of specific relationships is spun" (Vaggione 2001: 57). The articulation of artistic initiative through constraints becomes particularly essential when facing the open sound world of electronic music. These constraints can be applied to the materials, tools, and techniques to be employed in a composition process. Another constraint can be applied to the time allocated to the open-ended exploration of these materials, tools, and techniques. Ultimately, the seemingly arbitrary rules that govern the composer's plan, whether it stems from a procedure, narrative, or data set, help delimit the possible actions to a subset of all the creative routes that the electronic medium presents, narrowing the relationship between the medium and the composer down to the poietic thread unique to the piece at hand.

The Esthesic Thread

The human mind processes sensory input on the basis of what has already been experienced (Demers 2010: 50). The auditory system constantly absorbs new information from its surroundings and compares it to stored experiences (Truax 1984: 26). The stored experience of a sound shares a highly correlated perceptual space with the actual experience of the sound itself (Gygi, Kidd, and Watson 2007: 853). This is why our existing knowledge of likely sequences of sounds aids our auditory recognition (Gygi, Kidd, and Watson 2004: 1262). Through such interactions between memory and experience, the structured environments that we coevolve with establish a context for our auditory experiences (Windsor 2000: 20). As a result, our past experiences impact our appraisal of future encounters with auditory phenomena (Tajadura-Jiménez and Västfjäll 2008: 68).

We develop cognitive representations of auditory phenomena as elements of the meaningful events occurring in our daily environments; these representations are collective in terms of their relevance to the observer's membership to a community of experiences (Dubois, Guastavino, and Raimbault 2006: 869). When a listener is acclimated to a particular auditory environment, these learned representations establish a context for prediction and surprise (Huron 2006: 36). This quality of cognitive representations offers an insight into how connotations, as Meyer refers to them, can be more amenable to decoding in the context of electronic music. When our mental catalog of musical experiences fails to guide us through a piece of electronic music, the mind resorts to a more general catalog of experiences: a lack of musical reference conjures up a profusion of other kinds of references. The esthesic complexity of electronic music matches that of environmental sounds. The experience of meaning in electronic music is therefore in essential harmony with how we derive meaning from everyday experiences (Kendall 2010: 73).

The tendency of a listener to make sense of a sound in relation to their environment expands beyond the recognition of featural similarities between environmental sounds and electronic music; it also facilitates the attribution of contextual meaning to stimuli as the human mind concocts realistic spaces from abstract sounds (Wishart 1996: 146). Even the synthesized sounds in electronic music can evoke references in the listener's mind since those too can convey environmental information (Emmerson 1986: 26; Gaver 1993b: 290; Windsor 2000: 17). Smalley refers to this reflex of the human mind as *source bonding*, which can occur in response to "the most abstract of works" since bonding play is an inherent behavior of human perception. We seek supposed causes for sounds and therefore relate sounds to one another based on the assumption of a shared origin. The extrinsic links to the sounding world outside makes the intrinsic elements of a musical work meaningful as a cultural construct (Smalley 1997: 110).

However, even environmental sounds can elude semantic associations. In her categorization of everyday soundscapes, the cognitive scientist Danièle Dubois identifies *amorphous sequences* as those wherein specific events cannot be isolated. In a study on the identification of city noises, the participants tended to rely on acoustic properties such as intensity or pitch to distinguish between amorphous sequences (Dubois 2000: 48). This observation is in agreement with the results of an experiment conducted by the psychologist Catherine Guastavino, who found that the participants commonly used simple adjectives relating to physical attributes of the acoustic signal to describe *ambient noises*, wherein individual sound sources are not clearly identifiable (2007: 55). Similarly, Özcan and van Egmond argue that the featural aspects of sound become more salient when the recognition process fails to identify a source (2007: 199). These findings explain how some sounds in electronic music conform to neither musical nor environmental expectations, drawing the listeners to the low-level perceptual aspects of the musical experience.

In the next chapter, I will take a more practical approach to the esthesic thread when I introduce the details and results of a study that explores how listeners draw meaning from their experience of an electronic music piece. This will not only reinforce some of the ideas discussed in this section but also open up further discussions into the semantic processes that inform a listener's interpretation of this experience.

3

A Study on Listening Imagination

In this chapter, I will present an exploratory listening study on the experience of electronic music. This study was conducted internationally with eighty participants over the course of four years. It was aimed at gathering a detailed account of how a listener's experience of an electronic music piece operates on perceptual, cognitive, and affective levels, and elaborating a taxonomy of the common concepts activated in listeners' minds when listening to such a piece. I will first provide details of the five works of electronic music that were used as stimuli in the study. In addition to discussing the aesthetic and conceptual considerations that went into the composition of these works, I will give an in-depth account of the formal and technical implementations of these works, including descriptions of sound design strategies, production techniques, and hardware and software tools employed in their composition. I will then give a comprehensive report on the participants and the study procedure. Finally, I will present the results of the study and the various analytical methods applied to these results. In the following chapters, I will use these to formulate various theoretical constructs specific to the listening experience of electronic music. But before we go into the details of the study, let me provide some context with existing research that adopts a similarly experimental approach to the analysis of the electronic music experience.

A Cognitive Approach

The cognitive revolution of the 1950s brought about new perspectives on the study of the human mind. In psychology, the revolution took the form of a counter movement against behaviorism (Miller 2003: 141): striving to go beyond the study of observable behavior, cognitive psychologists began to investigate how representations of sensory stimuli are formed in the human mind and how these representations are processed by various mental faculties. In its rejection of the prevailing divide between mind and matter, a key premise of the cognitive revolution was to leverage concepts of information, computation, and feedback to ground the mental world in physical reality (Pinker 2003: 31). Psychology was not the only field to be greatly influenced by this revolution. In fact, the main impetus for this movement was its interdisciplinary disposition. The psychologist George Miller lists six disciplines that spearheaded the cognitive revolution: psychology, linguistics, neuroscience, computer science, anthropology, and philosophy (2003: 143). Unifying these fields through theoretical models, cognitive science thrives on the diversity of methods and perspectives that different disciplines contribute to the study of the human mind (Thagard 2014).

The study presented in this chapter adopts a cognitive science framework to investigate our perception of electronic music. It employs psychology and linguistics in its design, computational techniques in the analysis of the results, and philosophy to contextualize these results within the aesthetics and histories of electronic music. While the study is not actively engaged in neuroscientific inquiry, prior research in this domain has informed the formulation of the theoretical constructs presented later in the book. Furthermore, existing studies in anthropology and biomusicology that are referenced throughout the book illuminate evolutionary and cultural links between musical behavior in humans and our perception of electronic music. Overall, the study adopts a human-centered approach to the investigation of how we perceive electronic music, grounding theoretical perspectives in human experience through experimentation.

Experimental Studies on Electronic Music

The fields of psychoacoustics and music psychology deal with sound and music not simply as quantifiable phenomena but as objects of human perception. While the modes of inquiry central to these fields have existed since the ancient times (Yost 2015), both psychoacoustics and music psychology gained significant traction in the twentieth century in light of the aforementioned shifts in the scientific paradigm (Gjerdingen 2013: 694). As the cognitive revolution expanded into the study of auditory cognition in the latter half of the century (see Bartlett 1977; VanDerveer 1979), experimental studies on the perception and appraisal of music also began to proliferate.

Analytical perspectives on the experience of electronic music are almost as old as the genre itself. When Pierre Schaeffer formulated musique concrète in the 1940s, he did so by defining theoretical constructs, such as sound objects and listening modes, which can be viewed as proto-analytical devices. Today, Schaeffer's theories are still central to many analytical debates on electronic music. However, the post-Schaefferian era starting in the 1970s has produced its own array of analytical systems for the study of the electronic music experience (Demers 2010: 14). An extensive body of research has been devoted to the discourse on the ontological and experiential characteristics of electronic music with novel standpoints informed by semantics, perception, aesthetics, philosophy, and technology (see Chowning 1971; Emmerson 1986; Wishart 1996; Smalley 1997, Field 2000; Windsor 2000; Simoni 2006).

In human-centered research, user studies offer phenomenological insight into the subjective ways in which we experience the world around us (Stienstra 2015). A correlate of this approach in sound and music research is listening experiments where participants are asked to report on their experience of an auditory stimulus, whether this may be a brief audio sample or an entire piece of music. Although this form of inquiry is not unprecedented in the domain of electronic music, it is "the exception rather than the rule" (Landy 2007: 39). The number of listener-based studies focusing exclusively on electronic music is outstripped by the body of work on the cognition of instrumental music.

In one of the earliest studies of this kind, the musicologist Vladimír Karbusický surveyed 4,500 listeners about the notions and images that came to their minds while listening to three different excerpts from Herbert Eimert's *Epitaph for Aikichi Kuboyama*. The study involved interviews and questionnaires carried out with a diverse population representing a wide range of social status and age. Karbusický (1969) remarks that unlike the "stylized" sounds of conventional instruments, the new and unusual sounds of electronic music evoked in the listeners extramusical concepts that are much closer to their everyday experiences. For many participants, electronic sounds represented tokens of the modern world rather than autonomous sound structures, prompting the musicologist to devise a taxonomy of elicited notions consisting of catastrophes, strange sounds, and cosmic happenings. Karbusický also found that jazz enthusiasts had a much more positive attitude toward electronic music than folklore adherents, who rarely associated their experience with notions of joy (1969: 38).

In another early example, the music researcher Michael Bridger carried out an analysis of electronic music based on experience reports from groups of listeners (1989: 147). Using short excerpts from five works of electroacoustic music that make significant use of the human voice, Bridger first administered repeated listening sessions to acquaint the participants with the stimuli. In doing so, Bridger (1993) aimed to explore the kinds of imagery evoked by a piece and how a listener's appreciation of a piece is affected by their familiarity with it. Each listening session was followed by a discussion, where Bridger annotated listener comments on a waveform print of the piece. This method, although lacking statistical control, as Bridger points out, yielded several interesting insights such as the listeners' attentiveness to the human voice and the sounds of conventional musical instruments (1989: 159). Bridger also observed that spatial movement of sounds commonly created a positive response among the listeners even when a listener's overall reaction to a piece was not favorable. Interestingly, Bridger found that children had a generally more positive response to electroacoustic music than adults.

The researcher François Delalande was also an early adopter of the experimental approach in his investigation of listening behaviors in electronic music. In one study, Delalande and the composer Jean-Christophe Thomas invited eight subjects with varying degrees of experience with electronic music to listen to a short movement from Pierre Henry's *Sommeil* (Delalande 1998: 24). After the listening session, Thomas surveyed the subjects in an informal interview format. Delalande describes that, rather than focusing on the similarities between what the listeners thought they heard, they looked for consistencies among the listening behaviors they adopted (23). Even with the limited number of participants, Delalande was able to observe parallels between the listener testimonies. This allowed him to identify a set of listening behaviors, which he then utilized as a framework to analyze *Sommeil*. The behaviors that Delalande identified are taxonomic listening, wherein the listener develops a global synopsis of the work; emphatic listening, where the listener focuses on the sensations that the work elicits; and figurative listening, in which the listener associates sounds with living and moving beings. When pointing out the limitations of their study, Delalande acknowledges the lack of a systematic approach in the analysis of the listener reports and the insufficiency of the number of participants to draw statistical conclusions. On the other hand, Delalande remarks that a more empirical approach aimed at collecting simpler responses via questionnaires or segmentation tasks would not have yielded the same kind of rich feedback, which ultimately allowed him to identify the listening behaviors. As a general comment about these types of studies, Delalande characterizes verbal testimonies as a relatively reliable account of the listening experience, albeit affected by the "self-image of the listener" (25), implying that it is inevitable for the listener to project their own convictions while verbalizing their thoughts about a musical work.

Following a similar approach to the use of listener reports for musical analysis, the composer Andra McCartney carried out a listening study during the latter half of the 1990s focusing on Hildegard Westerkamp's soundscape music (McCartney 1999). In these studies, the participants were asked to listen to a soundscape piece and respond to a follow-up survey. The participants were asked to express their opinions about musical structure, imagery, memories, and places that the piece evoked. She then collated the listener reports to establish a framework for her analysis of the piece. In her analyses, McCartney contextualizes the results of the study by employing an "open interpretation" of the listener responses; the composer expresses that this decision was influenced by her desire to maintain an interpretational diversity in her analyses (198). Rather than drawing common conclusions about the perception of soundscape music, she uses the listener responses to form a dialogue about the individual works by juxtaposing perspectives from different disciplines (e.g., women's studies, ethnomusicology, music composition).

Rob Weale and Leigh Landy's Intention Reception Project investigates the relationship between the composer's intentions and the listener's experience in order to identify factors that affect appreciation in electroacoustic music (Landy 2007: 44). To achieve this, the researchers formulated a study that starts off with questionnaires directed at the composers whose pieces would be used in the listening experiments. This is followed by three listening sessions involving a real-time questionnaire, where the listeners are asked to list any thoughts, images, and ideas that come to mind as they listen to the piece. In the first listening session, the participants are given no information about the piece. Right after this session, they are given a directed questionnaire, where they are asked more specific questions about the narrative context of the piece and what the composer's intentions might have been. In the following two sessions, the listeners are given increasing amounts of information about the pieces before they respond to the same surveys as in the first listening session. The results reveal that when inexperienced listeners are provided with dramaturgical information about a piece, they are able to use this information to guide themselves through parts of the music that are problematic in terms of access and appreciation (Weale 2006: 196). Ultimately, Landy states that the Intention Reception Project demonstrates that the potential public interest in works of "sound organization" is greatly underrated (Landy 2007: 53).

Building upon Landy and Weale's project, the composer Andrew Hill used "empirical data collection" (2013: 43) to investigate audience reactions to electroacoustic audiovisual music. During various phases of this study, audience members were asked to respond to qualitative questionnaires that are based on those used in the Intention Reception Project (45). These surveys were intended to collect information about the material properties of an audiovisual work, the audience's semantic and emotional responses to this work, and whether the audience members wanted to keep experiencing the piece or gain contextual information about it (45). Hill argues that such contextual information can diminish the "interpretative potential of the work" (43). Although the mimetic material was initially found to be a barrier to the audience's aesthetic interpretation of a work, consecutive phases of Hill's study revealed that it was the contextualization of such material within the discourse of the work that had a more prominent impact on interpretation.

In another example, the composer Adam Basanta (2013) carried out an online survey with twenty-one participants to examine the language that listeners use while they reflect upon electroacoustic music. The stimuli for the study were short musical excerpts consisting of real-world field recordings and synthetic pitched material. Each participant was instructed to listen to an excerpt and provide affective, imaginal, and other kinds of responses in written format. From these responses, Basanta identified common concepts that the listeners employed when describing their experiences. For instance, most listeners positioned themselves as external entities in relation to the musical excerpts. They often identified place images that were motivated by spatial cues or sonic metaphors. Furthermore, ecological and cultural references were found to play a role in how listeners negotiated the meaning of a musical phrase. Basanta also notes that while the inexperienced listeners tended to focus on bodily associations, the experienced listeners adopted a more analytical perspective toward the musical excerpts.

More recently, the composer Mary Simoni (2018) carried out a study on audience perception of algorithmic music. The study uses reception theory to understand how listeners decode algorithmic music and how cognitive and affective factors influence the decoding process. Eight participants with no prior experience with algorithmic music were asked to write down their unstructured observations while listening to four different works of algorithmic music. Before a second listening session, the participants were given detailed information about the works. The listening session was followed by two surveys, one immediately after listening and another a month later. A discourse analysis was applied to the real-time survey responses to identify and classify cognitive-affective responses. The results indicate that while repeated encounters with a piece do improve a listener's ability to decode the composer's intentions, these may not necessarily contribute to aesthetic appreciation. Moreover, Simoni observed a prominence of affective responses when the subjects were given information to help decode the works, whereas the absence of such information resulted in a tendency towards cognitive responses.

"Talking about Music Is Like …"

As seen in these examples, and in many other projects presented throughout the book, the use of verbal feedback is a common method in studies on auditory perception. The saying that "writing about music is like dancing about architecture" has been attributed to numerous musicians from Frank Zappa to Thelonious Monk. Regardless of its origin, the sentiment of this quote is relevant here: once a listener begins to verbalize their thoughts about a musical experience, their responses become susceptible to influence from what Delalande refers to as the imprint of the listener's self-image. In other words, as we translate our experience of a musical work into rational thought, it becomes influenced by our lived experiences, expectations, and mood, among other factors. In McCartney's (1999) study, for instance, the participants' fields of expertise were shown to affect their verbal interpretation of soundscape pieces.

Not unlike the studies discussed earlier, the current study collects verbal feedback in the form of written and typed responses from participants. While "talking about music" can be problematic, the cognitive idiosyncrasies of electronic music discussed in Chapter 2 cast a different light on the relationship between one's musical experience and how they articulate this experience through semantic associations. Experimental evidence suggests that there is a substantial overlap between the neural networks dealing with verbal and nonverbal semantic information (Cummings et al. 2006: 92). Another experimental study conducted by Orgs et al. reveals similarities in the conceptual processing of verbal and nonverbal stimuli (2006: 267). Highlighting a strong conceptual relationship between sounds and spoken words, numerous studies on environmental sounds have demonstrated the facilitatory effect of everyday sounds on the retrieval of semantically related words (Ballas 1993; Van Petten and Rheinfelder 1995; Guastavino 2007; Gygi, Kidd, and Watson 2007; Özcan 2008). That is to say, a kind of music, the vocabulary of which has been thrown open to all sounds via the electronic medium, can be more amenable to talking about.

Stimuli

Five complete works of electronic music were used in the listening study. Four of these, namely *Birdfish, Element Yon, Christmas 2013*, and *Diegese*, are my own works. These four pieces can be found on anilcamci.com/ccem. The fifth one is a piece by Curtis Roads titled *Touche pas*. In this section, I will give an in-depth report on the various materials, techniques, and compositional intents involved in the composition of my works. I will also provide details of *Touche pas* in the form of program notes and personal communications that I had with Roads.

The variety of the materials and techniques employed in my works is rooted in my artistic practice, on the one hand, and the goals of the research presented here, on the other. In that regard, each piece was composed with a hypothesis built into it. It could be argued that such hypotheses are inherent to every artistic process: the artist develops assumptions about the experiences their work might elicit. Even with generative or stochastic pieces, where the artist may have minimal control over the unfolding of the work, such assumptions exist. The spectators approach the experience of an artwork with a similar set of assumptions. What's essential to point out here is that these two sets of assumptions—namely, that of the artist and that of the audience—can be diametrically opposed. In laying out the compositional plans for the pieces discussed here, my goal is to reveal my artistic perspective, which I do not view as the ground truth. When we discuss the results of the study later on, we will clearly see the ways in which my assumptions overlap or conflict with those of the listeners. More interestingly, there will be overlaps and conflicts between the listener reports, which will give us a glimpse into the cognitive processes at play while listening to electronic music.

As we discussed in Chapter 1, I refer to a broad range of practices when I use the phrase *electronic music*. Sure enough, other composers' works that I reference throughout this book come from unmistakably different styles. This is because electronic music can be created in a number of ways: it can be performed with hardware or software instruments; it can be montaged from recorded or synthesized sounds; it can be based on chance operations or computational algorithms. Not only are these just a few of the methods and materials made available by the electronic medium, but they are also often used in conjunction. We first observed an amalgamation of production techniques in the early years of electronic music when composers began to transcend the stylistic boundaries of their studios. Aleatoric processes made their way into deterministic styles and synthesized signals got intermixed with concrète sounds. As technology advanced, production techniques became homogenized in ways that foster new practices.

This is not to say that modern electronic music practices are indistinguishable from one another. On the contrary, such combinations of styles and techniques have paved the way for countless microgenres ranging from glitch hop and vaporwave to tech trance and chiptune. However, when an artist resorts to a limited set of tools today, their choice is more likely to be rooted in aesthetics rather than technical limitations such as those that informed the work of early composers of electronic music. Nevertheless, the works presented here utilize a diverse range of production techniques. While some of the pieces were composed entirely out of synthesized sounds, others incorporated live recordings. Some of the pieces were based on predetermined narratives, while others grew out of process-based composition. Procedural production methods were used in tandem with stochastic techniques. Parts of these works were created with generative systems, while others were performed with instruments. This variety was not necessarily intended to investigate the experiential qualities of these techniques in isolation, but to achieve a gestalt of techniques afforded by the modern electronic medium.

The decision to work with my own works was motivated by several factors. Firstly, as the creator of these works, I possess a firsthand knowledge of not only the tools and techniques but also the artistic intents and strategies involved in their composition. This creates a unique opportunity to construct a dialogue between the artist and the listener on account of the level of detail I can unpack about these works. In her overview of analytical approaches to electronic music, the musicologist Laura Zattra argues that one of the main challenges in this domain is the multiplicity of the ways in which a piece can be represented (e.g., symbolic notation, time-stamped events, data pertaining to sound synthesis) and the lack of analyses that leverage a heterogeneous documentation of all these dimensions. With this study, I hope to address such limitations by imparting an extensive amount of information about the pieces and adopting several analytical approaches to the interpretation of the listener reactions. Furthermore, since the study is designed to gauge listeners' unmediated and spontaneous reactions to electronic music, I avoided the use of canonical works to minimize the possibility of participants having prior exposure to the study pieces.[1] This is also why the study instructions given to the participants did not include any information about the works.

Later in this chapter, I will offer an exhaustive and transparent account of the study method in terms of how the listener reports were collected and analyzed. I refer to these reports throughout the book to exemplify certain experiential qualities of electronic music and to motivate new perspectives and models toward the analysis of this kind of music. I believe that the novelty of the insights I am able to draw as the creator of these works outweighs the noise that my involvement with this material might introduce to the data. Nevertheless, to diminish the impact of a confirmation bias on my part, the study method and the analysis of the results were peer-reviewed both during the study itself and through various conference and journal articles published since (see Çamcı 2012b, 2013, 2016; Çamcı and Meelberg 2016; Çamcı and Özcan 2018).

A Note about Fixed Music

The current study focuses exclusively on fixed—rather than live-performed—electronic music for several reasons. The focus on fixed pieces made it possible to design a much more controlled (and technically feasible) study that was repeated many times across multiple countries. More importantly, a live performance entails unique experiential qualities. Besides the potential variations in the musical outcome from one performance to another, there are many variables that can influence an audience's reaction to a performance. For instance, the presence of a performer can affect how the audience associates sonic figures with actions or objects. While the current study is not concerned with acousmatic experiences in the Schaefferian sense, it deals with the listener's unmediated experience with sounds. This is why the unique circumstances of a live electronic

---

[1] I was able to skirt this issue with Roads's *Touche pas* because this particular piece had been performed only a few times prior to the study. The piece became publicly available years after the study as part of Roads's 2019 album *Flicker Tone Pulse*.

music performance was left outside the scope of this book. There are, however, numerous studies that focus on such circumstances (Bahn 2001; Oliveros 2005; d'Escriván 2006; Jensenius 2007; Ciciliani 2014), including those of my own (Çamcı, Çakmak, and Forbes 2019; Çamcı, Vilaplana, and Wang 2020).

Birdfish (2012, 4'40")

Birdfish was composed between October 2010 and February 2012 as the third entry in a tetralogy on evolutionary phenomena. The works in this collection (in the order that they adhere to in the narrative arc of the tetralogy) are *Let the World End* (2012), a fixed piece; *Nautik* (2011), a live electronic performance; *Birdfish* (2012), the fixed piece discussed here; and *Insectarium* (2013), an interactive audiovisual installation. While these pieces belong to a preconceived sequence, they vary substantially in terms of material, form, and style. The individual parts of the tetralogy are envisioned as standalone works and are not intended to be performed in a single program. Besides the overall theme of evolution, these works share sound design research and techniques. For instance, the design of water textures in *Birdfish* are informed by the performance of *Nautik*, where I synthesize various forms of liquid sounds over the course of 12 minutes. The phase modulation techniques developed to create the biophonic sounds in *Birdfish* were later used in *Let the World End* to form the central motif of the piece. Finally, my research on aerial and avian sounds for the composition of *Birdfish* was utilized to design the interactive sounds of *Insectarium*. *Birdfish* narrates a metamorphosis of underwater beings into avian creatures. Throughout the piece, synthesized sounds of water, amphibians, fish, and birds are introduced at varying degrees of intelligibility. These elements intermittently morph into a recurring melodic leitmotif that marks the transitions between sections. The individual sound elements, their motion trajectories, and the resonant spaces they move within are designed to instigate clear representations of the objects and creatures that populate the universe of the piece.

Sound Design

Despite its representational intents, *Birdfish* does not consist of any recorded sounds. All of the sounds in this piece were generated using a combination of pulsar synthesis, granular synthesis, and frequency modulation (FM) synthesis. These were then processed with various modulated delays, filters, and reverberation. The granular techniques used for creating the textural elements in the piece involved stochastic processes that helped execute the organic transitions between different timbres. Once I created the foreground and background elements via several iterations of the said processes, I micro-montaged these on the timeline of the piece across various spatial layers.

Frequency modulation is a key characteristic of bird vocalizations (Stowell and Plumbley 2014). The acoustic properties of birdsong are manipulated by multiple modulations at different rates. Furthermore, the amplitude envelope of a chirp shares a similar contour as that of its pitch envelope, giving the bird vocalizations their typical temporal unfolding where the pitch rapidly glides up. A simple method of synthesizing bird vocalizations involves the use of a ramp envelope to control both the amplitude and the frequency of a sine wave; gradual increases in volume and pitch are followed by a rapid roll-off in both parameters. An additional low-frequency oscillation (LFO) can be used to modulate the frequency of the gliding pitch, adding complexity to the vocalization through vibrato. The bird sounds in *Birdfish* were synthesized following a similar principle, as shown in Figure 3.1. The output of an envelope generator controls the frequency of a carrier oscillator exponentially to create a rapid increase in pitch in each chirp. The same envelope controls the amplitude of the output linearly to give the chirps a dynamic contour. Finally, a second oscillator linearly modulates the carrier frequency.

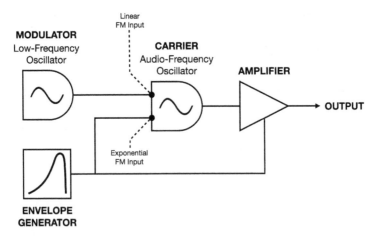

Figure 3.1 Simple bird-vocalization synthesis using an envelope generator and two oscillators.

The water sounds in *Birdfish* were achieved through a variety of techniques. Applying an LFO to the amplitude of a white noise source is often used to imitate wind and wave sounds. The LFO can also be applied to the center frequency of a low-pass filter to color the white noise as its amplitude oscillates with the opening and closing of the filter. This is an efficient and relatively convincing sound design technique for simulating the experience of listening to a large body of water from a distance. The composer Suzanne Ciani, for instance, uses this technique in her performances to create an ocean scenery that immerses the audience. For water sounds observed up close, more microscopic variations need to be added to the design. In *Birdfish*, these are achieved through vocoding and granulation. In the simple vocoder implementation, a sound source is passed through a series of band-pass filters that divide the spectrum of the input signal into a number of frequency bins. The energy in each bin is encoded with amplitude followers that generate control signals that are then used to resynthesize the input signal by modulating the amplitudes of band-pass filters applied to a broadband noise source. Through this process, the pitch content of an input signal can be modified in real time to add melodic and harmonic properties that are not present in the original signal. Although the first musical applications of the vocoder dates back to the early days of the electronic music studio,[2] its extensive use by bands like Kraftwerk and Daft Punk made it a staple of modern electronic music. In *Birdfish*, a white noise source, which exhibits a uniform but randomized energy across the frequency spectrum, is fed into a vocoder to modulate the band-pass filters applied to a sawtooth oscillator. The randomness inherent to the white noise signal manifests itself as artifacts that resemble bubbling or streaming water.

Form

Birdfish consists of two movements that themselves are composed of two and three sections, respectively (Figure 3.2). The 38-second exposition acquaints the listener with the sonic palette of the piece with obfuscated expositions of the bird, water, and other creature sounds as well as intermittent melodic materials. The

---

[2] Homer Dudley, who developed the vocoder at Bell Labs in the 1930s, gave a demonstration of his device in Germany in 1948. In attendance was Werner Meyer-Eppler, whose work on synthetic sounds is considered to have led to the creation of the Cologne studio at the Westdeutscher Rundfunk in 1951 (Luening 1964: 97).

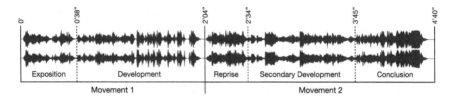

Figure 3.2 Overall structure of *Birdfish* shown on a waveform display of the piece.

first instance of a leitmotif, which will recur throughout the piece, marks the end of this section with an E-flat resolution. The following development section, which lasts until the 2′04″ mark, contains more concrete incarnations of the water, fish, and amphibian sounds. This section builds up on the representationality of its sound elements, yet at the same time introduces more abstract objects such as brief glides of discernibly synthesized sounds. This section also reveals the spatial design of the piece by introducing reverberant fields, low-frequency rumbles, and higher-frequency gestures that traverse the implied environment stereophonically. These spaces are designed to enhance the contextual coherence of the previously introduced representational sounds, essentially gluing together the foreground elements to the overall narrative of the piece. The leitmotif marks the end of the section, this time in D-flat. Throughout this section, rests play a fundamental role in articulating not only gestures but also the spaces around them as they draw the listener's attention to the reverberant tails of the sounds preceding them.

The second movement begins with a 30-second section of fast-paced mutations between the actors introduced so far (i.e., birds, water-like elements, amphibians, and abstract structures) and ends with an evaded cadence-like reincarnation of the leitmotif which *dissolves* into a reverberant field as the pitch slips downward from D. With this, the collage of bird sounds sinks into the background, opening up several compositional prospects: while the gestural pace of this background matches that of the previous sections, the figure elements are now pushed away spatially to distance the listener from the main texture that has persisted throughout the piece so far. The shifting of the figure gestures to the ground sets a stage for a new foreground, which I populate with a more sporadic distribution of sounds that evolve over time from abstract pitched elements back into creature sounds. Further into this section, the *sunken* gestures of bird and amphibian sounds swing in and out of reverberation to reclaim their foreground roles ephemerally. The reverberation and the low-frequency pulses in this section are reminiscent of the first movement but are much more articulated to create a spatial tension for what is to come. An increasing number of representational sounds come out of and go back into the reverberant field. Following the most articulated submersion into reverberation between 3'35" and 3'45", a brief reprise of the textural density of the first movement lends itself to a final resolution involving a climactic build-up across various dimensions of the piece including spatial, spectral, and textural densities, and representational variety. The effect I wanted to achieve here was to shift the perspective from the third person to the first person and have the piece *wash over* the listeners, as if they were among the creatures that were striving to transcend the surface of the ocean. A final repetition of the leitmotif concludes the piece in D-flat.

In addition to such narrative uses of reverberation, certain gestures are expanded spatially using audio decorrelation between the left and right channels. In this technique, a monophonic signal is doubled and sent to the left and right channels individually. Then one these channels is delayed by less than 50 milliseconds. Such a delay falls within the Haas window, ensuring that the listener does not hear two separate instances of the sound as the auditory system fuses the dry and the delayed signals into one. The delay effectively replicates the interaural time difference, which is a sound localization cue that our auditory system uses to pinpoint the location of a sound. The perceived effect is a stereo widening of the monophonic signal.

Element Yon (2011, 3'45")

Element Yon was composed concurrently with *Birdfish* during the nine-month period between October 2010 and June 2011. The two pieces share similar gestural, temporal, and spatial organizations. However, such similarities between the two pieces are obscured by the marked differences in the sonic material. Unlike the sounds of *Birdfish*, which were designed to be representative of living creatures and environmental phenomena, the sounds of *Element Yon* were performed with a subtractive synthesizer and maintain the pure characteristics of simple waveforms. In many ways, this choice of material casts *Element Yon* as an abstract counterpart to *Birdfish*. Moreover, the composition of the synthesizer performances into the eventual fixed piece does not follow a narrative structure, mirroring the abstract quality of the sound material at a formal level.

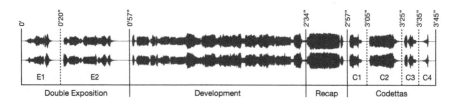

Figure 3.3 Overall structure of *Element Yon* shown on a waveform display of the piece.

Sound Design

Just like *Birdfish*, *Element Yon* is composed entirely of generated sounds, with FM playing a central role in the design of many of its sonic elements. The primary sound sources for *Element Yon* are the low-pass and band-pass filters of a Korg MS-20 synthesizer with their resonances turned up to put the filters into self-oscillation. Through live performances supplemented by a modulation matrix involving the envelope generators and LFOs found in this synthesizer, the self-oscillating filters interact with each other in complex ways. As the two oscillations travel between the extremes of the audible frequency range, the beat frequencies occurring between them create timbral variations throughout the piece. Much like in *Birdfish*, the use of rests plays a significant role in the sound design of gestural articulations in *Element Yon*. Reverberation effects are used not only to blend between moments of sound and silence but also to contextualize the abstract sonic structures in unrealistic spaces.

Some of the modulations applied to the filter frequencies are the same as those used for the synthesis of avian vocalizations in *Birdfish*. As a result, certain gestures in *Element Yon* verge on the organic sounds in *Birdfish* in terms of their structural complexity. In contrast with the aforementioned harmonic evasions in *Birdfish*, which transpire in a representational sound world, the evasions in *Element Yon* are those of representationality within an abstract context.

Form

Element Yon adopts a pseudo-sonata form consisting of exposition, development, recap, and coda sections (Figure 3.3). Although this form does not share the binary model of *Birdfish* at the global scale, the macro- and gesture-level structures of *Element Yon* are similar to those of *Birdfish* in terms of temporal extent and the use of rests in between the structures.

The piece begins with an exposition consisting of two sections: The 20-second opening introduces the vocabulary of the piece with phrases that operate at different time scales. This is followed by a 33-second section that further establishes the frequency range of the piece. Throughout the exposition, the listener is acquainted with the gestural characteristics of *Element Yon* and the sound qualities they will encounter throughout the piece.

During the development section between 0'57" and 2'57", the piece expands spatially with the use of the same audio decorrelation technique employed in *Birdfish*. The section starts off by exploiting the low end of the frequency spectrum, which is maintained for most of the section in the style of a pedal point. In the higher range of the spectrum, decorrelated gestures go in and out of unison with the low end. Throughout this section, the tonal structure of the piece evolves from atonality into an implied Phrygian dominant scale in A#.

The so-called codettas between 2'57" and 3'05", 3'07" and 3'25", 3'27" and 3'35", and finally 3'36" and 3'45" are the "four elements" which give the piece its name.[3] These four segments revisit themes from the development in the form of self-contained units that do not display structural relationships among each other. Although the second segment consists of a harmonic resolution in the vaguely implied key of the piece, the four segments in combination are intended to obfuscate any motivic closures to the piece, in the manner of *moment form*. Karlheinz Stockhausen, who first conceived the moment form as a structural unit for his piece *Kontakte*, describes that a moment is "not merely regarded as the consequence of the previous one and the prelude to the coming one, but as something individual, independent and centered in itself, capable of existing on its own" (Kramer 1978: 179). Along a similar line, the four elements that conclude *Element Yon* serve independent musical functions rather than an overarching goal.

Christmas 2013 (2011, 2'16")

A Christmas song for a world that no longer exists, as crooned by the anti-Santa strolling a wasteland formerly known as Earth: A repurposing of Tin Men and The Telephone's celebrated Christmas album. (Çamcı 2012a, concert program notes)

---

[3] "Yon" is a Romanization of the Japanese word 四, which means "four."

In 2011, the jazz trio Tin Men and the Telephone offered me uncompressed stereo mix-downs of the tracks from their then-upcoming album *The Very Last Christmas*, which consists of avant-garde interpretations of famous Christmas songs. Inspired by the ill-conceived prophecies of the world's end in 2012, I wanted to use this material to create an electronic Christmas song set in a postapocalyptic world, about a year after the supposed demise of humankind. Instead of recounting a particular story, the piece portrays a situational narrative centered on the theme of *future nostalgia*, where a hypothetical future is presented as the past. The composition of *Christmas 2013* was completed in a relatively short period of two months.

Sound Design

The instrumentation of the trio consists of piano, double bass, and drums. For *Christmas 2013*, I had decided early on not to use any material besides the tracks I received from the trio, constraining the sonic vocabulary of the piece to the mix-downs of the instrumental recordings and their time-domain and frequency-domain manipulations. During the preparation phase, I listened to the recordings repeatedly to extract audio samples that display interesting transient and spectral qualities. Once I sorted out the sound material, I began to experiment with processes that I could apply to this material. To support the theme of future nostalgia, I decided to maintain references to the original recordings. At the same time, I aimed to strike a balance between sounds that are electronic and precise and those that have lo-fi and organic qualities. I then used feedback delays, filters, reverberation, and micro-montaging to give the selected samples a variety of qualities ranging from futuristic to antiquated.

The piece is structured around a temporal unfolding marked by a staccato gesture. The recurrences of this gesture in different timbres and on different time scales establish an obscure rhythm that operates at various levels of form throughout the piece. The spatiotemporal organization of these staccato sounds reveals a collective behavior based on spatial causality, akin to flocking, to give the listener the impression of observing a landscape populated with animate objects. Throughout the piece, the staccato gesture manifests itself in various ways, often fed into a pitch-shifting delay line where the frequency of each repetition gets gradually shifted down. Combined with spatial movement, this effect was intended to create a sense of decay and withdrawal. This spatial

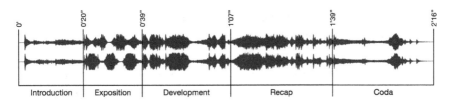

Figure 3.4 Overall structure of *Christmas 2013* shown on a waveform display of the piece.

behavior also plays an important role in articulating space as the repeating staccato sounds travel around the listener throughout the piece.

Filtering and reverberation serve multiple functions in *Christmas 2013*. While low-pass filters were often used to create lo-fi versions of certain gestures, they were also combined with reverb effects to simulate localization cues; these cues define a multilayered spatial environment, where the sound elements move not only laterally on the stereo panorama but also backward and forward. In addition to articulating space, the reverberation also serves to create ambient textures that establish a tonal thread throughout the piece. In a couple of instances, a low-frequency sound is fed into a reverb with a long decay to create a rumbling effect that gives the listener a sense of tectonic movement within a vast space.

Form

Christmas 2013 is loosely based on the sonata form (Figure 3.4). The piece opens with the first instance of the staccato sound in the style of an orchestral stab. This is followed by a quotation of the Christmas carol *Silent Night*. This part is kept true to its form in the original recording apart from added reverberation. This is intended to prime the listener with an instrumental reference that they can easily recognize as a song, if not as the specific carol that it is. The reverberated melody is then layered with a harmonically related texture, gradually expanding the ambient space of the piece. Toward the end of the introduction section, the sound of the trio performing *Silent Night* becomes overtaken by this texture.

The exposition section starts with a differently pitched version of the staccato sound, this time fed into a delay line with medium feedback and a downward pitch shift. A spatial play of impulses takes over the foreground and begins to immerse the listener in an expansive environment. The exposition is concluded with a gesture wherein the repeating impulses pan around the listener with their pitch falling below the audible range, creating an overall sense of a downward spiral.

The start of the development section is marked by another instance of the staccato sound that articulates the widest space so far. This is followed by a sparse distribution of brief impulses intermixed with instrument sounds within this reverberated space. The sporadic appearances of the piano sounds are placed among their highly processed counterparts in an ornamental fashion. The narrative function of these appearances is to remind the listeners of the *now*, in a future from where they are looking back. The now therefore becomes an object of nostalgia through the contrast between an unsettling postapocalyptic landscape and pleasant memories of familiar sounds. A low-end rumble further establishes the vastness of the implied environment. Another low-end rumble starting at 1′07″ marks the beginning of a section that recapitulates some of the themes introduced so far with original gestures contrasted with their low-pass-filtered versions, demonstrating a wide frequency range.

The coda section starts with an electronic line performing an atonal melody. This line is gradually overtaken by an ambient texture similar to that from the introduction. The texture crescendoes back to the clearly identifiable sound of the piano now playing a progression fabricated from processed recordings to compliment the tonal and temporal threads of the piece. This progression concludes the piece with a harmonic resolution.

Diegese (2013, 1′54″)

Diegese is a short-form piece composed between June 2012 and February 2013. Some of its sound material was composed to illustrate the role of diegesis in electronic music during a talk that I gave at the Toronto Electroacoustic Symposium in 2012. In this talk, I blended my live speech with recordings of my speech, synthesized sounds, and sounds of acoustic instruments to highlight the ways in which musical material can be given narrative roles that are external and internal to the implied universe of a story. Over the months following this presentation, I continued to experiment with these ideas using the nonverbal sound material from my talk to compose a fixed electronic music piece. A report on these and other experimentations are offered in my article "Diegesis as a Semantic Paradigm for Electronic Music," published in the Canadian Electroacoustic Community's journal *eContact!* in May 2013.

With *Diegese*, I specifically wanted to explore the placement of musical quotations within electronic music as *diegetic actors*. We will cover the concept of diegesis more extensively in Chapter 5. Briefly put, diegesis is used to articulate the many layers of a narrative and how the elements of the said narrative are situated relative to one another. For instance, a diegetic sound in film is one that the characters of the story can hear, whereas a prime example for non-diegetic sound is the score of the film, which is only audible to the spectators of the film. Therefore, a diegetic actor is an element that belongs to the story as opposed to being external to it like the score. The idea of making a musical excerpt a diegetic actor in a broader piece implies that the excerpt is encapsulated in the narrative of the said piece. To materialize this idea, which I had previously experimented with in *Birdfish* and *Christmas 2013*, I incorporated two musical quotations into the composition of *Diegese*. The quotations are blended into the structural flow of the piece rather than being juxtaposed with the remainder of the sonic textures.

Sound Design

The sound sources in *Diegese* comprise of a Moog Slim Phatty synthesizer, a custom software synthesizer, and a piano recording. These sources are then processed with spectral and spatial effects before they are montaged on the timeline of a DAW.

The first of the two aforementioned quotations is from Curtis Roads's piece *Touche pas*, which was also used in the listening study. The quoted texture is Roads's homage to his teacher Morton Subotnick's piece *Touch* (Roads 2009, concert program notes). In *Touch*, percussive sounds looped at different intervals create a chorus of asynchronous rhythms. Roads describes that he stumbled upon a similar texture while he was experimenting with the granulation of an impulse. In my own recreation of this texture, I created a custom software, where a sine wave is windowed by an envelope, the length of which is stochastically modulated within a range of 20 to 30 milliseconds. The frequency of the sine wave is randomized in a range of 80 to 2500 Hz before each windowing. The enveloped signal is then randomly sent to one of three different delay lines, the delay times for which are randomized in a range of 10 to 100 milliseconds. Since each delay line is set to high feedback, they function like loopers that approximate the behavior of a granular synthesizer at shorter delay times. This way, the delay lines create concurrent streams that are then spatialized on the stereo panorama. This results in a texture comprising grains that pulsate aynchronously in the style of *Touch* and *Touche pas*.

The second quotation is a phrase from Ludwig van Beethoven's *Piano Sonata No. 27 in E minor* (Opus 90). The choice of instrument here sets a clear parallel with *Christmas 2013*, where the piano sound was similarly a recognizable element of the sound palette. The main difference between the two quotations is that while *Christmas 2013* incorporates an external musical form as a whole (i.e., a Christmas carol), *Diegese* uses the piano as a decontextualized quotation which can be associated with the instrument but not the musical form it is taken from.

In *Diegese*, I utilize a compositional construct that I refer to as a *Sonic Rube Goldberg Machine*. I had previously used this construct in my 2007 piece *Shadowbands* and my 2009 piece *Hajime*. A Rube Goldberg Machine is a convoluted contraption that is designed to perform a simple task (Olsen and Nelson 2017: 104). A famous example of a Rube Goldberg Machine is found in the opening scene of the classic science-fiction film *Back to the Future*, where we see a complicated mechanical system involving clocks, switches, a toaster, and a robotic arm performing tasks akin those fulfilled by modern home automation systems. A Rube Goldberg Machine is characterized by the causal relationships between the individual parts that make up the machine: a rolling ball sets off a kettle, the vapor from the kettle floats a balloon toward a needle, the pressure from the popping balloon topples the first one in a series of domino pieces. A Sonic Rube Goldberg Machine emulates the physical causality between these consecutive actions in the auditory domain: a sound object mimicking the behavior of a bouncing ball as it moves through the stereo sound field collides with another object, setting it on its own motion trajectory, and so forth.

The pairing of the two quotations in *Diegese* is based on the similarities between the metaphorical movements that these quotations exhibit. The looping of concurrent impulses at varying time intervals in *Touche pas* is reminiscent of ping-pong balls being dropped on the floor. The descending arpeggio in the quotation from Beethoven's Opus 90 displays a similar downward motion, almost as if the musical phrase is falling down. The accelerando arpeggiation of the Moog synthesizer is designed to imply a similar metaphor. Throughout

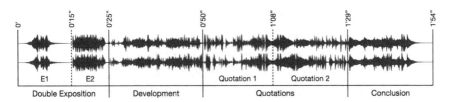

Figure 3.5 Overall structure of *Diegese* shown on a waveform display of the piece.

the piece, these gestures are juxtaposed with upward glissandi and reversed reverb tails to create the illusion of physical causality in the style of a Sonic Rube Goldberg Machine; objects gain potential as they move up before they eventually fall down.

Form

The opening of the piece until the 0′15″ mark clues the listener in about the quotations that are the focal points of the piece without fully introducing them (Figure 3.5). The reversed reverb tail of the piano recording activates the looping of a single impulse. Although the first exposition offers a brief excerpt from the arpeggiated synthesizer, this sound is fully fleshed out in the second exposition.

The development section defines a spatial environment animated with various transient gestures. These are in fact reversed and slowed-down versions of what would be introduced as the first quotation at the 0′50″ mark. However, when presented in this form, the causal relationship between the gestures becomes easier to follow, albeit in reverse. This way, the whole development section functions as an upward build-up toward the quotations section. Concurrent to this causal interplay of gestures is an ambient layer that further articulates the spatial extent of the environment with reverberant tails responding to pitched impulses. Although this layer is situated much further from the listener, the impulses operate within the same causal network, functioning as distant echoes of the gestures that unfold in closer proximity to the listener. As this second layer reveals the outer bounds of the sonic space, a third and even closer layer is populated with decorrelated granular clicks. The lack of room effects on these clicks paired with the stereo separation effect, similar to what was used in *Birdfish* and *Element Yon*, gives them an in-the-head feeling.

The following section introduces the two quotations in succession. Halfway through the first quotation, the granular texture from *Touche pas* is brought into the same spatial environment as that of the development, now layered with slowed-down and reversed versions of itself. A gesture micro-montaged from this quotation activates the reversed reverb tail from the opening of the piece. This time, however, the tail reveals itself to stem from a piano recording, setting in motion the second quotation. As the piano arpeggio descends, it also becomes intermixed with the first quotation, leading into a reprise that brings back the development themes. The piece is concluded with a variation on the introduction.

Touche pas (by Curtis Roads, 2009, 5'30")

In the program notes for the debut performance of *Touche pas*, Roads wrote,

> In April 2008, I was experimenting with the granulation of a brief sound file when I noticed that its opening looping sonic texture was reminiscent of the classic electronic music work *Touch* (1969) by my former teacher Morton Subotnick. I know this work well, having produced and remastered it in 1986 for the Wergo label. In homage, I decided to dedicate this work, whose French title means "touch not," to Morton Subotnick. The title proved apt, as after months of work, *Touche pas* sounds nothing like *Touch*. (Roads 2009, concert program notes)

In my personal conversations with Roads, he described that he did not plan the final structure of the piece in advance. Instead, he relied on intuitive explorations of the sound material and extensive editing to give the piece its eventual form:

> *Touche pas* is an example of how a combination of generative synthesis techniques (i.e., interactively-controlled granulation) with detailed surgical intervention can bear fruit. Granulation is an efficient means of spawning reams of new material out of a tiny fragment of sound. The original source material for this piece was a 3-second fragment consisting of 21 impulses. Using the EmissionControl program, I was able to telescope this fragment by a factor of 200 into a 10-minute texture full of variation.
>
> In the months that followed, I extracted the most salient structures within this flux. In effect I divided the original sound file into dozens of smaller files, while relentlessly discarding redundant material. In several stages of construction, I mixed these files into a new ordering while simultaneously

designing *microfigures* at key moments. Microfigures are patterns of transient sound particles in rapid succession, for example, a series of echoes ascending by an interval of a perfect fifth. In many cases I inserted the smallest of particles—impulses lasting less than 1 ms—to construct a kind of transient sparkle. These extremely short clicks, at the limit of human auditory perception, are rarely arranged artistically. Yet they have the wonderful quality of pricking the ears of the listener, making one aware of the ever fleeting present instant in a direct and physical way. The density of the microfigures challenges the listener to focus on the figures inside the phrases on the micro time scale. Because of this density, this work lasts less than six minutes, divided into two parts. (Roads 2016, personal communication)

Study Design

The study employs a between-subjects design, where each participant listens and responds to only one of the five pieces. The study is comprised of two stages that complement each other in capturing a comprehensive representation of the participant's experience of an electronic music piece. The first stage, namely the general impressions task, is aimed at gathering an overall account of the participant's interpretation of a complete piece of electronic music. This section allows the participant to evaluate the piece holistically and provide their post-hoc impressions in free form. The second stage, namely the real-time descriptors task, collects the momentary impressions that surface in the participant's mind as they listen to the same piece in its entirety for a second time. This stage is aimed at collecting gesture-level descriptors that pertain to perceptual, cognitive, and affective processes. While the first stage allows for an uninterrupted listening experience, the second stage facilitates a continuous and immediate reporting of the mental associations evoked by the piece, with each descriptor mapped onto the timeline of the relevant musical piece. In combination, the two stages of the study collect both contextual and momentary concepts that a piece might evoke.

Although the use of short audio excerpts in listening studies can be useful for researchers to focus on isolated aspects of the listening experience, this approach yields responses that do not account for the entire context of a piece and can fail to capture the dynamic aspects of musical experience that unfold over a larger timescale (Chapin et al. 2010: 2). The motivation behind using complete works in the current study was to allow the communication between the artist and the listener to transpire to its intended extent.

> ## A Note about Experiment Bias
>
> Experimental studies, such as the one described here, are prone to artifacts that can influence the outcome of the study. One such artifact, called the *observer expectancy effect*, occurs when the experimenter subconsciously communicates their expectations to the participant, encouraging responses that support the hypothesis of the study (Zoble and Lehman 1969: 357). Another possible artifact is called *demand characteristics*, where a participant forms an opinion about the aim of the study and gives responses that reinforce or nullify the hypothesis (Thomas 2010). These artifacts are inevitable to some extent, simply because any verbal or written communication between the experimenter and the participant will establish a framework of expectations. These effects can be mitigated by maintaining a consistent level of bias across all instances of the study through well-rehearsed instruction routines.
>
> A common practice in experimental studies on music perception is the upfront stipulation of a task to be performed either during or immediately after a listening session. In the former case, where the participants are asked to take notes or respond to a questionnaire while they listen to a piece of music, the usual music listening experience is interfered with. Although this method can prove effective to gauge momentary reactions to a piece, it requires the participant to be repeatedly distracted from the act of listening. In the latter model, in which a task is described prior to a listening session that will be followed by a survey, the listening experience remains uninterrupted. However, this model nevertheless requires the listener to assume a certain stance toward the piece. It is worth noting that a biased stance toward a piece can never be fully eliminated from experimental studies, owing to the listeners' awareness that they are participating in a study. However, giving the participant a task prior to a listening session will inevitably encourage them to put conscious effort into fulfilling the task as the piece plays, which also implies a notable alteration of the listening experience. The current study aims to mitigate the effects of such biases by way of adopting a novel combination of these two approaches.

Preliminary Studies

The experimental design of the current study is informed by two preliminary studies. The first one was a general impressions study conducted with eight people in December 2011. All eight participants listened to *Christmas 2013* in a concert hall and subsequently reported their general impressions in written form. The second preliminary study was a pilot of the broader study described here, consisting of both a general impressions task and a real-time descriptors task. This was conducted between October 2011 and February 2012 with twelve participants, who listened to an early version of *Birdfish* in individually administered sessions that were conducted using the custom software that would be used in the main study. These first twelve instances of the real-time descriptors task helped identify necessary improvements to the initial study design ranging from interface refinements to broadening of the collected data. Most prominently, whereas the preliminary study only tracked descriptor submission times, the eventual system in the main study tracked the moment at which a participant started typing a descriptor and then again when they submitted it. This change made it possible to observe with greater accuracy the exact time in the piece when a particular descriptor would emerge. More importantly, this implementation provided information regarding the time a participant spent between starting to type a descriptor and submitting it, which allowed for meaningful interpolations in several cases. The results of both preliminary studies will be reported in the following section. The methods used to evaluate these results informed the analysis of the main study. Furthermore, the descriptor categories extracted from the preliminary study served as a basis for the categorization of the descriptors in the main study.

Participants

Sixty participants from thirteen different nationalities took part in the main study between May 2012 and July 2014. Twenty-three (38.3 percent) of the participants were female, while thirty-seven (61.6 percent) were male. The average age of the participants was 28.78, with ages ranging from 21 to 61. The pool of participants included professional musicians, music hobbyists, composers, sonic arts and sound engineering students, as well as twenty-two participants (36.6 percent) who described themselves as having no musical background. The primary motivation behind working with a large group of participants with diverse backgrounds in terms of age, nationality, and prior engagement with electronic music was to alleviate the effects of potential biases rooted in culture and experience.

Based on the timing data obtained from the exercise section, all participants proved to be capable of typing five-letter words in less than 1.5 seconds, indicating a typing speed of 40 words per minute or faster, which is significantly higher than the average number of real-time descriptors per piece. This implies that typing proficiency did not constitute a major performance bottleneck for the participants.

Although all of the participants described themselves as English speakers, they were told that they could respond in their native language if or when they preferred to do so. Two participants used Turkish for both the written and the typed sections of the study. Two other participants typed their real-time descriptors in Turkish despite having written their general impressions in English. A few participants typed occasional descriptors in their native languages (e.g., *kraai*, meaning "crow" in Dutch; and *canicas*, meaning "marbles" in Spanish) but responded in English throughout the rest of the study. Any non-English response was translated to English prior to analysis.

Setup

The study was conducted using a custom browser-based software programmed in HTML, CSS, and PHP. The software has multiple views with simple interfaces, as seen in Figures 3.6 through 3.9. The interfaces consist of text-input fields, drop-down lists, and buttons for controlling audio playback and moving between views. The descriptors submitted via the text-input fields are logged in an SQL database with timestamps that represent the time at which a participant starts typing a descriptor and the time when they submit the descriptor. The browser-based software is run on either a laptop or a desktop computer with a QWERTY keyboard. The listening tasks are carried out with closed-back or semi-closed-back stereo monitoring headphones.

Procedure

Each study takes approximately twenty minutes. The pieces are rotated between participants to achieve a random allocation with an equal number of instances for each piece. Verbal instructions are provided prior to each section of the study. The study procedure involves an initial listening session, a general impressions task, a real-time descriptors exercise, and a real-time descriptors task.

Initial Listening Session

The participant is seated in front of a computer displaying the software interface seen in Figure 3.6. After a brief description of the interface, it is explained that,

Figure 3.6 Initial Listening Session software interface.

once the participant presses the play button, they will start listening to a piece of music, which will be played in its entirety without interruptions. No information regarding the piece (e.g., title, duration, composer name) is disclosed to the participant. They are asked to approach it like any piece of music to the best of their ability. The participant is then given a pair of headphones to listen to one of the five works of electronic music used in the study.

General Impressions Task

Once the initial listening session is completed, the subject is instructed to write on a piece of paper their general impressions as to anything they might have felt or imagined, or anything that came to their mind while they were listening to the piece. This instruction is intended to cover a wide range of mental activations that could represent perceptual, cognitive, and affective processes. It is explained to the participant that they could write freely in whatever form and to whatever extent they prefer without a time constraint. Once the participant indicates that they have completed the general impressions task, they are asked to return to the computer and press the continue button on the software interface where they left off. In the following page, they are presented with a digital form, as seen in Figure 3.7, where they can input their personal details. Once this is completed, the participant proceeds to the exercise section for the second part of the study.

Real-Time Descriptors Exercise

In the real-time descriptors exercise, the participant is presented with the interface seen in Figure 3.8. It is explained that once the participant presses the play button, they will hear the voice recording of a person reading a piece of text. They are instructed to pick random words from this text, type them in the text box, and hit the return key on the keyboard to submit the words

84   *The Cognitive Continuum of Electronic Music*

Figure 3.7 Participant information form.

Figure 3.8 Real-Time Descriptors Exercise software interface.

one at a time. It is stated that once they press play, the cursor would move into the text box automatically, allowing them to type into the text box. It is also explained that once they hit the return key, the text box would be cleared, and the cursor would move back to its initial position. This design ensures that the participant can secure their hands over the keyboard during the real-time descriptors task without having to use the mouse to input descriptors.

The main purpose of the exercise is to acquaint the participant with the software and hardware layout of the study platform. By filling out the personal information form and completing the exercise section, they get familiarized

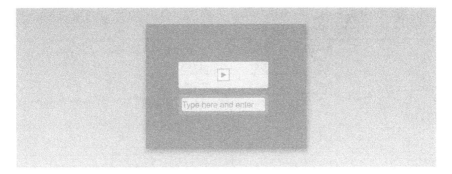

Figure 3.9 Real-Time Descriptors Task software interface.

with the input mechanism. The exercise section also shows if there are any connection problems between the software and the SQL database where the entries are logged. The recording used in the exercise is 30 seconds long. However, if the participant experiences a problem during the exercise, it can be played back again. Once the exercise is completed, the participant is asked to press the continue button to proceed.

Real-Time Descriptors Task

In this section, the participant uses the interface seen in Figure 3.9, which is almost identical to that from the exercise section, to complete a real-time free association task. Prior to this task, it is described to the participant that once they press the play button, the piece which they previously listened to would be played back a second time in its entirety. It is explained that, in this section, they are expected to submit descriptors as to what they might feel, imagine, or think, the moment such descriptors come to their mind. The participants are advised to be relaxed and spontaneous as they type these descriptors. The interface for this section is designed to encourage this kind of spontaneity. Although the database underlying the interface supports extended text input, the single-line text box is intended to deter the participant from spending too much time on preparing prose-form entries and, as a result, losing their active engagement with the piece. The participant is also asked to disregard any typing errors and submit their descriptors as soon as they finish typing them.

Table 3.1 Total and Average Numbers of Real-Time Descriptors (RTDs) Submitted per Piece, per Participant, and per Minute per Participant

	Birdfish	*Element Yon*	*Christmas 2013*	*Diegese*	*Touche pas*
Piece duration	4'40"	3'45"	2'16"	1'54"	5'30"
Total number of RTDs	334	170	198	161	339
Average number of RTDs per participant	27.83	14.16	16.5	13.41	28.25
Average number of RTDs per minute per participant	5.96	3.77	7.27	7.05	5.13

Results

The general impressions were reported in one or a combination of various forms, including a list of words, a list of sentences, prose, and drawing. Although no time constraints were specified for this section, most participants spent between five and ten minutes to complete the general impressions task. Table 3.1 provides an overview of the number of descriptors submitted in total for each piece in the real-time descriptors task. Although the vast majority of the descriptors were in the form of single words or two-word noun phrases, there were longer descriptors as well. The longest descriptor submitted was a ten-word sentence.

Data Visualizations

In auditory perception studies that deal with short audio samples, statistical representations of data can be sufficient to draw conclusions from the results. However, the use of complete works of electronic music in the current study made it necessary to incorporate the temporality of the listening experience into the analysis. Given the sheer number of descriptors submitted by the participants, it became apparent early on that custom tools for data visualization would be needed for a meaningful evaluation of how the descriptors relate to a piece both within and between participants. For the comparative analysis of the real-time descriptors, I developed two interactive visualization systems.

Figure 3.10 Single-timeline dynamic visualization of real-time descriptors.

Single-Timeline Dynamic Visualization

The single-timeline dynamic visualization places all the descriptors submitted in response to a piece on a single timeline. This allows for a sequential analysis of the entries and provides a compiled overview of descriptors from multiple participants. However, given the number of real-time descriptors submitted for each piece as seen in Table 3.1, it was impossible to make each descriptor readable when placed on a static timeline. To overcome this issue, I developed a dynamic visualization that reacts to the passage of time by highlighting the descriptors submitted at a given moment in the piece.

All descriptors pertaining to a piece are placed on this timeline in a vertically cascading pattern, as seen in Figure 3.10, with a vertical line drawn from the descriptor to its exact point on the timeline. Pressing the space bar on the keyboard starts the playback of the piece. As the playback proceeds, the descriptors that were submitted in the vicinity of the current moment in the piece dynamically expand as seen in the figure. In order to maintain the temporal relevance between the piece and a given descriptor, the visual placement of each word is based on the time at which the participant started typing the descriptor. An entry begins to expand as the elapsed time approaches its point on the timeline and reaches its most expanded form at the time when the typing of the descriptor had begun. After this moment, the descriptor starts to shrink back to its initial state on the timeline. This dynamic behavior of the visualization makes it possible to view all the descriptors on a single timeline and establishes a sense of context for each descriptor.

Figure 3.11 Multiple-timeline visualization of real-time descriptors by two participants.

The elapsed portion of the piece is displayed in a darker color on the timeline. Clicking on the timeline allows for jumping to different moments in the piece. The dynamic visualization also responds to these jumps by expanding the descriptors at the clicked point. This makes it possible to explore the relationship between a particular gesture in the piece and the multitude of descriptors submitted by various participants around the moment where the said gesture happens.

Multiple-Timeline Visualization

To perform contextual analyses of descriptors within and between participants, it became necessary to visualize all the descriptors by individual participants on separate timelines. This resulted in multiple timelines to be visualized concurrently, as seen in Figure 3.11. Similar to the single-timeline dynamic visualization, this visualization is also capable of playing back the audio file relevant to the data that are being visualized. In this case, however, the visualization is static except for the progress bar. It is likewise possible to jump to different points in the piece by clicking on the timeline; doing so updates the progress bar for all the timelines, effectively highlighting the correlations between participants. This visualization is also useful when performing per-participant contextual analyses of individual descriptors.

Analysis Methods

Given the extent and variety of the data obtained from the study, the interpretation of the results required the use of various analysis techniques

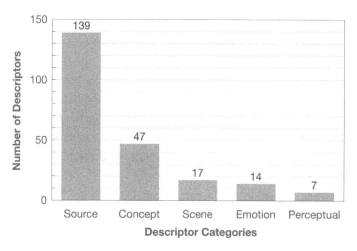

Figure 3.12 Categorization of the descriptors gathered from the second preliminary study conducted with twelve participants listening to an early version *Birdfish*.

including descriptor categorization, correspondence analysis, and discourse analysis, as well as comparative analyses of the general impressions and the real-time descriptors using the aforementioned visualizations.

Categorization of Descriptors

In order to analyze the descriptors gathered from the real-time free association task, a categorization was imposed upon the data, following the model of many previous studies on auditory perception (e.g., Ballas 1993; Marcell et al. 2000; Guastavino 2007; Gygi, Kidd, and Watson 2007; Özcan 2008). In the second preliminary study, an iterative process of thematic analysis was applied to the data to produce a set of descriptor categories. Once the emergent categories were determined, the categorical membership of each real-time descriptor was assessed through forced-choice categorization. As seen in Figure 3.12, the categories derived from the preliminary study were *source, concept, scene, emotion,* and *perceptual descriptors*.

To analyze the data from the current study, all of the 1,202 real-time descriptors were initially grouped under these five categories. If a descriptor consisted of multiple words or noun phrases, it was split up into its constituents, which were then categorized individually. For instance, the descriptor "computers underwater" was broken into "computers" and "underwater." When categorizing ambiguous descriptors, three cues were utilized: the moment in the piece where the descriptor occurred, the context of the descriptor (i.e., the adjacent descriptors), and the musical background of the listener. Five descriptors whose categorical correspondence could not be determined due to either obscurity (e.g., "but then again") or over-generality (e.g., "sound") have been left out of the categorization. After several iterations of the categorization process, it became apparent that some of the categories derived from the preliminary study were either too broad and had to be split up into subcategories, or they were insufficient to represent some of the descriptors and therefore required new categories to be devised.

Upon further evaluations of the categorical distributions, a list of labels that adequately represented the data was established. This final list of categories addresses the various stages of meaning attribution, including perception, recognition, and identification (Özcan 2008: 18), as well as processes of affective appraisal. The said list includes the following descriptor categories:

- Source Descriptors (SD), subcategorized into:
    o Object Descriptors
    o Action Descriptors
    o Musical Descriptors
- Concept Descriptors
- Location Descriptors
- Affective Descriptors (AD), subcategorized into:
    o Emotion Descriptors
    o Appraisal Descriptors
    o Quality Descriptors
- Perceptual Descriptors (PD), subcategorized into:
    o Auditory Descriptors
    o Featural Descriptors
- Meta Descriptors
- Onomatopoeia

The Source Descriptor category includes descriptors that can broadly be prefixed by the phrase "sound of." The three subcategories are Object Descriptors (e.g., "water," "telephone," "frogs," "wind"), Action Descriptors (e.g., "breathing," "explosion," "scratching," "bouncing"), and Musical Descriptors (e.g., "guitar," "lullaby," "pop band"). Objects and actions can refer to both animate and inanimate entities. Similarly, musical descriptors are objects of both animate (e.g.,

"Mozart") and inanimate (e.g., "percussion") nature. The choice of separating musical descriptors from object descriptors stems from the significant number of relevant entries; this will also facilitate a discussion into the idea of *music within music* in Chapter 5.

The Concept Descriptor category includes such descriptors as "waiting," "lights," "transition," and "summer." As seen in these examples, concept descriptors can also be objects or actions; however, they do not refer to sounding objects or phenomena in themselves. On the other hand, these descriptors might refer to concepts that imply such phenomena, as in "war," "Chinese," and "science fiction."

Location Descriptors refer to imagined spaces different from the one inhabited by the listener (e.g., "jungle," "underwater," "cave," "hallway"). A location descriptor can also indicate imaginary spatial attributes, as in "distant," or merely imply an imagined yet unspecified environment, as in "space" and "outdoors."

Affective Descriptors are grouped into three subcategories. Emotion Descriptors define feelings that relate to the listener's experience, such as "curious," "stress," "relief," and "fear". Appraisal Descriptors such as "nice," "cool," "lovely," and "great" are often followed by a source descriptor, as in "nice piano" or "cool low." These descriptors denote a listener's basic appraisal of certain components of the piece on a binary basis (i.e., good or bad). Quality Descriptors such as "weird," "familiar," "exciting," and "mellow" are affective traits that the listener attributes to an external object, as in "relaxing rhythm." Therefore, the difference between emotion and quality descriptor categories is that while the former denotes a feeling of the listener, the latter describes a feeling of an object.

Perceptual Descriptors are grouped into two subcategories. Auditory Descriptors denote perceptual qualities of the sound, such as "bass," "silence," "fade in," and "pan". Featural Descriptors denote nonauditory perceptual qualities of the imagined objects, as in "wide (room)," "small (impacts)," "deep (cave)," and "dark (forest)."

Meta Descriptors refer to the material being of the piece in itself and not the experience of it (e.g., "(great) opening," "want more bass," "pause," "end"). Such descriptors can also refer to form and technique (e.g., "counterpoint," "granular," "motif," "pitch-shifter").

The Onomatopoeia category includes a small number of descriptors such as "boooooom," "ding," and "hummm."

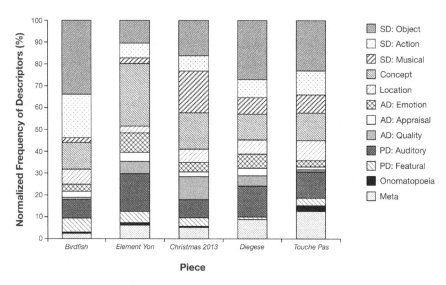

Figure 3.13 Categorical distribution of real-time descriptors by piece.

Figure 3.13 shows the frequency distribution of each category per piece. These categorical distributions in themselves are already revealing of some of the experiential differences between the five pieces. The reader of this book is invited to listen to these works and compare the results with their own impressions. I will use these distributions in the coming chapters to draw conclusions both within and across the pieces.

Comparative Analysis

The two-stage design of the study allowed for various comparative analyses. Firstly, the general impressions and real-time descriptors were compared within participants to observe the semantic correspondence between the two stages of the study. Furthermore, the general impressions were compared between participants in the discourse analysis described below. For the comparative analysis of the real-time descriptors between participants, I used the custom visualization software described earlier.

The data from the two stages of the study displayed significant similarities within most participants. The basic themes and concepts appearing in the general impressions of a participant were commonly apparent in their real-time descriptors as well. As a result, the within-participant comparative analyses allowed the use of general impressions to contextualize certain real-time descriptors in the broader framework of a piece and the themes that the listener inferred from it. The correspondence between the general impressions and the real-time descriptors also made it possible to identify which part of a piece a general impression might be referring to by looking at the time stamp for the real-time descriptor associated with this general impression.

Correspondence Analysis

Correspondence analysis is a statistical method for visualizing the relationships between the layers of a frequency distribution matrix. The rows and columns of a two-way contingency table are displayed as points in a low-dimensional space with the aim of maintaining a global view of the data in a way that facilitates its interpretation (Lee 1996: 65). The application of this technique to the current data offers a representation of the categorization results as a distribution on a two-dimensional graph in relation to their frequency of occurrence in each piece. This representation contextualizes the pieces among each other and the descriptor categories as seen in Figure 3.14. The five pieces are marked on the correspondence graph with an "×," while the descriptor categories are marked with a "☐."

Discourse Analysis

The free-form general impressions allowed for various analyses, some of which we have discussed above. Another analysis method applied to this data was discourse analysis, wherein general impressions expressed in a range of formats (e.g., prose, list, drawing) are split into "meaningful sections" (Özcan 2008). For instance, the sentence "it reminded me of marbles falling" is reduced to the words "marbles" and "falling." This way, a list of keywords that represent a participant's general impressions is generated. These keywords, in return, are grouped between participants by semantic similarity and evaluated on the bases of their rate of occurrence. The tables that follow provide these keywords as well as a breakdown of the formats used in the general impressions per piece. Multiple formats used by a single participant (e.g., a word list and a drawing) were counted separately. Table 3.2 shows the keywords that occurred multiple times in the first preliminary study where eight participants provided their general impressions of *Christmas 2013*.

Tables 3.3 through 3.7 show the keywords for the pieces in the main study. Here, a similar analysis has also been applied to the real-time descriptors. For a real-time descriptor to be evaluated in the discourse analysis, it needed to be repeated by at least three separate participants. The number of occurrences for each word is denoted next to it within parentheses. Real-time descriptors that show semantic similarities were collated under a single descriptor, with the total number of occurrence indicated in parentheses. The collated descriptors are also provided at the bottom right of the table in a separate list.

Figure 3.14 Correspondence analysis between pieces and descriptor categories.

Table 3.2 Keywords Gathered from the General Impressions Provided in the First Preliminary Study Conducted with Eight Participants Listening to *Christmas 2013*

General Impressions (prose (4), sentence list (3), word list (2))
Christmas, [music, instruments, drums, piano, harmonium], [tonal thread, melody, cadence], [nostalgia, memory, 80s, childhood], [open, space, air], [flying, movement, distance], [slow, time], event, real

Table 3.3 Keywords Gathered from the Feedback for *Birdfish*

General Impressions (word list (5), sentence list (4), prose (3), drawing (1))

[water, bubble, splashing, sparkling, fluid, flow, liquid, waves, lake],
[living, creatures, animal, amphibian, bird],
[slimy, worm, snail, squishing, insect, swarm],
[alien, Zerg, StarCraft, sci-fi, Star Wars],
[high tech, robots, electronic],
[granulating, grinding],
[metallic, blades, gong],
[sense of space, cave, Efteling], bass, dialogue

Real-Time Descriptors (used by at least three participants)

water (12), underwater (3), bubble (3),
bird (7), flying (4),
creature (7), cat (4),
bass (6), big (5),
laser (4), war (3), metal (3),
bug (3),
(sense of) space (3),
mouth (3),
high frequency (3), small (3)

Collated Descriptors:

water, wet, liquid, fluid, something spilling;
bug, ant, swarm;
animal, creature;
mouth, eating;
bass, low frequency;
high frequency, high pitch

Table 3.4 Keywords Gathered from the Feedback for *Element Yon*

General Impressions (prose (6), word list (3), sentence list (3), drawing (1))

[wide spectrum, spectral, high frequency, low frequency, contrast],
[electronic, oscillators, synthetic, abstract],
[unpredictable, unstable, unclear, confusing, surprising, exciting],
[dangerous, scary, chaotic, argument],
[painful, irritating, exhausting, annoying], [relief, relieving, calm],
[slow movement, stable, still], silence,
[science fiction, Tron]

Real-Time Descriptors (used by at least three participants)

disturbing (4),
rest (4), relief (3),
sweep (4),
static (3),
conversation (3)
high (3), frequency (3),
pan (3),
repeat (3),
(outer) space (3),
cry (3),
I (3)

Collated Descriptors:

disturbing, painful, annoyed, irritating;
rest, freeze, wait, silence;
relief, release, peaceful;
conversation, communication, he's trying to tell us something;
cry, scream;
high, sharp;
frequency, tone, pitch;
sweep, slide, fall;
repeat, again;
space, spaceship;
pan, travel, move

Table 3.5 Keywords Gathered from the Feedback for *Christmas 2013*

General Impressions (prose (5), sentence list (5), word list (4), drawing (1))

[creepy, scary, thriller, paranormal activity], [nervous, stressed, anxiety],
[relaxing, happy, calm, relieved, hope, fairy-tale],
[melodic, piano, music, song, ballet, cliché, familiar],
[space, extraterrestrial, astronomy, science, computer],
water

Real-Time Descriptors (used by at least three participants)

music (9), piano (9), scary (5), machine (5), (sense of) space (4), suspense (4), storm (4), rumble (3), electronic (3), familiar (3), nice (3), bird (3)	*Collated Descriptors:* music, soundtrack, ballet, a dance, Mozart, jazz, Yes (band), Pink Floyd, Christmas song; electronic, electricity; scared, scary, Paranormal Activity (film), creepy, death; rumble, low drone; machine, robot, modem, matrix, inhumane; storm, thunder, turbulence; suspense, expectation, anticipation, waiting

Table 3.6 Keywords Gathered from the Feedback for *Diegese*

General Impressions (sentence list (6), prose (5), drawing (2), words (1))

[piano, keyboards, instrumental, song], [comfortable, pleasing, cool, happy],
[ball, ping pong balls, spherical, circular objects], [bouncing, drop, percussive, impulse],
[tiny organisms, insects],
[sands, grains],
[pond, liquid, water, waves, boiling, humid],
[imaginary, mysterious, science fiction, Alice in Wonderland]

Real-Time Descriptors (used by at least three participants)

piano (7), insect (5), ball (5), drop (3), door (3), dense (3), exciting (3), weird (3), bass (3), panorama (3)	*Collated Descriptors:* insect, bug; pinball, ping pong ball, ball, bubble; dense, complex; weird, uncomfortable, creepy

Table 3.7 Keywords Gathered from the Feedback for *Touche pas*

General Impressions (prose (6), sentence list (4), word list (3), drawing (2))

[marble, ball, bowling ball, circular, coin],
[bouncing, dripping, falling, breaking, impact, percussion, door (knocking)],
[granular, pieces, particles],
[convincing physicality, visual],
[calm, relaxing, meditative, relief],
[silence, pauses, ending],
[water, fluid],
[distant, far away, afar], sense of space,
material, panorama, motif, fun

Real-Time Descriptors (used by at least three participants)

ball (6),	*Collated Descriptors*:
again (5), repetition (4),	
water (5), drop (3),	ball, marbles;
percussive (4),	grain, granular;
(sense of) space (4),	percussive, gong, xylophone, woodblock;
bells (4),	reverse, rewind
reverse (4),	
grain (3)	

4

The Electronic Gesture

In this chapter, we will focus on the communication of meaning in electronic music by exploring the parallels between how we make sense of electronic music, on the one hand, and environmental sounds, on the other. We have already discussed the evolutionary disposition of the auditory system to identify sources for sounds and how this tendency can come into effect in response to most abstract of sounds. In such cases, our cognitive faculties turn to our catalog of experiences with everyday situations, wherein our interpretation of environmental sounds are tightly coupled with the events that they originate from. We will further unpack the relationship between events and environmental sounds by taking a closer look at mental representations, specifically those that pertain to our everyday experiences. This will help us identify the event units by which we experience electronic music and arrive at an idiomatic definition of gestures in electronic music. We will explore various aspects of the electronic gesture such as meaningfulness, causality, coexistence, and intentionality.

Events in the Environment

According to Lakoff and Johnson, we impose boundaries on physical phenomena to make them into discrete entities bounded by a surface much like our embodied selves; our experiences with physical objects provide us with a variety of ontological metaphors to view events as self-contained structures (2003: 26). Events are the units by which we make sense of what transpires around us. We encapsulate our experiences in time and space to derive useful information from our environment. The sun rises, the water boils, the flower blossoms, and the traffic light turns green. Multimodal stimuli originating from these events are picked up by our sensory mechanisms and processed by our cognitive faculties. We organize, fuse, and supplement sensations of different kinds to form intermodal associations (Gibson [1979] 1986: 245). Sound is often a component of the associated sensations that we gather from our surroundings. This is why the cognitive representations of acoustic phenomena exist in multimodal contexts (Dubois, Guastavino, and Raimbault 2006: 869). We engage with and interpret environmental sounds as an inherent component of the multisensory events we encounter in our daily lives.

Environmental Sounds

According to the psychologist Nancy VanDerveer, environmental sounds (1) are produced by real events, (2) are more complex than laboratory-generated sounds such as pure tones, (3) have meaning by virtue of indicating events in the environment, and (4) are not part of a communication system (1979: 17). Let's set aside the first proposition initially and focus on the other three. Environmental sounds indeed exhibit a higher level of structural complexity than laboratory-generated pure tones and pitched instruments. As we touched upon in Chapter 2, the frequency spectrum of an environmental sound rarely displays harmonic patterns. Furthermore, the dynamic contour of an environmental sound can be quite complex based on the physical phenomenon that it emanates from. This coupling between a sound and its source underscores the role of an environmental sound as a signifier that carries meaning. This is why the last item in VanDerveer's definition—that environmental sounds are not part of a communication system—is a necessary distinction to make: an environmental sound is not a symbolic representation of an event; the two are causally linked with each other. To illustrate this point, Ballas and Howard point out speech as an example of a sound that does belong to a communication system (1987: 97). A spoken word is only meaningful within the set of communicative conventions agreed upon by the speakers of the language that the word belongs to, whereas the meaning of an environmental sound is not tied to such conventions.

This same principle also distinguishes environmental sounds from musical sounds. In Chapter 2, we talked about certain aspects of musical appraisal that are informed by cultural conventions versus those that are rooted in biology. The culturally informed traits of our appreciation of music can be considered part of a communicative system—however subjective the communication itself might be. Having said that, the distinction between an environmental sound and a musical sound can be quite fluid. The transition from one to the other is a function of intentionality. The sound of a footstep is an environmental sound that is indicative of someone walking. In this case, the intention of the person who is walking is going from one place to another; the footsteps are a natural byproduct of this person's actions. If this person were to walk to the rhythm of a musical piece, the causal link between the event (i.e., walking) and the environmental sound (i.e., footstep sounds) is reversed. Their intentionality shifts from the act of walking to its acoustic outcome, at which point the footsteps can be considered musical sounds. For an onlooker, there may be no audible difference between the two activities, and therein lies the subjectivity of the communication. We will further delve into the role of intentionality in musical behavior in the next section. But before doing that, let's go back to the first item in VanDerveer's definition to explore the relationship between environmental sounds and "real events."

When an object of perception, such as the sound of an event, is treated as a sign, it assumes a semiotic meaning relative to the object it is associated with (Jekosch 2005: 199). The logician Charles Sanders Peirce groups signs into three categories based on their relationship with the objects they signify: *symbols*, *icons*, and *indexes* (Burks 1949). A symbol has an arbitrary relationship with the object based on conventions or rule systems that require learning. For instance, we need to learn that the word "happy" signifies the feeling of happiness since the semiotic link between the two is arbitrary. An *icon*, on the other hand, shares qualitative similarities with the object. This can be an image, metaphor, or diagram that indicates the signified object. Finally, an *index* is physically or causally related to the object. A mercury thermometer, for instance, signifies the air temperature by way of being governed by it. These signs exist in a conjoint space between the things that they denote and the mind that interprets them (Peirce 1885: 180).

Environmental sounds are in an indexical relationship with events. An event, in its simplest form, is "a thing that takes place" (*Oxford Dictionary of English*) and "a sound is news that something's happening" (Jenkins 1985: 117). This is why we process and categorize environmental sounds as meaningful events that provide relevant information about our environments (Guastavino 2007: 54) and why, in our daily lives, we listen to events rather than sounds themselves (Gaver 1993b: 285). This understanding is evidenced in several cognitive studies that highlight an intrinsic relationship between environmental sounds and events on the basis of source and action properties (Gygi, Kidd, and Watson 2007: 853; Brazil, Fernström, and Bowers 2009: 2). Marcell et al., who conducted an experiment on confrontation naming of environmental sounds, found that sounds in our daily environments primarily convey action and movement-related information (2000: 833), supporting the idea that the human perception is wired to go beyond acoustic patterns and recognize the events underlying these patterns (Ballas and Howard 1987: 97).

In these experimental studies on the cognition of environmental sounds, action and source properties are often found to be the most salient features that participants utilize to describe sound events. Action, in this context, is the process that an object goes through (e.g., shattering is an action a piece of glass can go through). While some researchers demarcate *source* as an object–action compound (e.g., shattering glass), others identify only the object as a source and reserve a separate category for actions. In Chapter 3, the descriptor categorization derived from the current study followed the latter model, where objects and actions were identified as separate source categories.

Models of Mental Representation

Our minds create mental representations of the objects we encounter in everyday life. We use these representations to make sense of our reality. Research in cognitive psychology offers various models for how such representations are formed. In one of those models, the psychologist Lawrence Barsalou proposes the concept of *perceptual symbols* as representations that underlie cognition (1999: 577). According to Barsalou, when we perceive an object, our selective attention extracts a subset of its features to be stored in long-term memory. This perceptual memory can serve a symbolic function, where it stands for referents in the world on later retrievals. Based on the assumption that perceptual symbols are immediate causes of imagery experience, the cognitive scientist Nigel Thomas (2010) situates Barsalou's theory of perceptual symbols closely with the concept of *mental imagery*, which he identifies as a quasi-perceptual reconstruction of a past perceptual experience in the absence of an external stimuli. This, however, does not imply that perceptual symbols are exclusive to visual representations since they can emerge in all modalities of experience including vision, audition, smell, taste, touch, action, emotion, and introspection (Pecher, Zeelenberg, and Barsalou 2003: 120).

The developmental psychologist Jean Piaget's model of schemas can be viewed as a precursor to the theory of perceptual symbols. In this model, various performances of the perceptual system become integrated and abstracted into systems of experience called *schemas*, which enable us to process objects of perception as signs (Jekosch 2005: 200). We utilize these schemas as systems of experience that help us parse objects and actions. When we experience an event, we store multimodal representations of this experience. Upon an encounter with a new event, we assimilate and accommodate "non-fitting data of perception" (204) and relate them to a previous experience of a similar event in an attempt to satisfy our innate mechanisms of anticipation. This tendency to accommodate non-fitting data helps explain why listeners are able to identify real-world referents to sounds that are not only synthesized but are synthesized without a poietic goal of instigating such references.

Much like in other modalities, mental representations of acoustic phenomena are concept-driven memories of previous perceptions (Dubois, Guastavino, and Raimbault 2006: 869). In these representations, the event-related meaning (i.e., semantic features) of a sound can be more salient than its physical attributes (Guastavino 2007: 60). For instance, acoustically dissimilar sounds produced by the same type of event are perceived to be more analogous to each other than acoustically similar sounds that originate from different events (Gygi, Kidd, and Watson 2007: 849). This perceived uniformity between sounds from similar sources helps listeners develop mental models of sound-producing events.

The composer Barry Truax's concept of *earwitness accounts* adopts an *acoustic communication* perspective toward the mediation of event-related information through memory processes. Acoustic communication investigates the relationship between sounds, the listener, and the environment by situating them as parts of a system rather than as isolated entities (Truax 1984: xxi). Studies in this field focus on the reciprocal relationship between an acoustic environment and its inhabitants (Çamcı and Erkan 2013: 20). This is why acoustic communication is inherently based on an *ecological approach*, which has been described by Gaver as leveraging patterns of information grounded in sources and environments, unlike traditional research methods that focus on primitive components of sounds such as loudness and frequency (1993a: 5). Truax distinguishes between the traditional and ecological approaches by characterizing them as the *energy transfer model* and the *communicational model*, respectively. While the former deals with sound as a physical phenomenon in isolation, the latter investigates the exchange of information and the cognitive processes that underlie this exchange (Truax 1984: 3, 9). Echoing VanDerveer's emphasis on meaningfulness of environmental sounds, Truax situates such sounds as mediators between the listener and the environment rather than mere vectors of energy transfer (11). This proposition, which highlights the role of sounds in building a relationship between an environment and its inhabitants, can also be aligned with Molino's esthesis-poiesis model that we discussed in Chapter 2. In that model, a directional hierarchy of communication from the composer to the listener was superseded by a neutral level called the trace, which acts as a mediator between the composer and the listener.

Affordances

An ecological approach to perception that is commonly applied to musical research (see Windsor 1995; Nussbaum 2007; Östersjö 2008) is the model of *affordances* formulated by the psychologist James Gibson. Gibson's studies on ecological perception stemmed from his experiments in aviation during the Second World War. Focusing mainly on an active observer's perception of their environment, Gibson postulated that the invariant features of a visual space provide pivotal information for perception; invariants are the attributes of an object that persist as the point of observation changes (Gibson [1979] 1986: 310). While Gibson's research on invariants is primarily grounded in the visual domain, his concept of affordances has been applied to other modalities of perception, including hearing.

According to Gibson, objects in an environment, by virtue of their invariant features, afford action possibilities relative to the perceiving organism. For instance, a terrestrial surface, given that it is flat, rigid, and sufficiently extended, affords for a human being the possibility to walk on it (Gibson [1979] 1986: 127). His main motivation to propose this seemingly straightforward idea is to oppose the prevailing models of perception that assume that we extract meaning from our sensory experiences by imposing mental structures upon the inherently chaotic stimuli that we gather from our environments. Gibson refutes this theory by suggesting that there are certain kinds of structured information in the form of invariants available in the environment even before the stimuli that carries this information reaches our perceptual systems. The kinds of structured information that an invariant affords are relative to the perceiving animal (Gibson 1966: 73). An object will therefore have different affordances for different perceivers: a stone, on account of its physical properties, affords the action possibility of *grabbing* for a human being, while it affords the action possibility of *climbing* for an ant.

One of the criticisms of Gibson's model is targeted at his epistemological stance on the roles of learning and memory in ecological perception. In his review of Gibson's seminal book from 1979, *The Ecological Approach to Visual*

Perception, the psychologist Bruce Goldstein argues that Gibson understates the role of learning in how we interpret the meaning of objects (1981: 193). Gibson proposes the idea of *perceptual knowing* to challenge the dichotomy between perception and cognition, which at the time was prevalent in the field of psychology. Relying on the concept of invariants, Gibson suggests that "perceptual seeing is an awareness of persisting structure" (Gibson [1979] 1986: 258) and knowledge of this structure already exists in the environment for a perceiver to obtain. Although Goldstein states otherwise, the role of learning is patently brought into discourse not only in Gibson's 1979 book but also in his earlier writings. Gibson deems it unquestionable that an infant has to learn to perceive by exploring the environment with all of his organs, "extending and refining his dimensions of sensitivity" (Gibson 1963: 15; 1966: 51, 285). The perceptual system matures with learning, allowing the information that it picks up to become more elaborate and precise as life goes on (Gibson [1979] 1986: 245). He argues that knowledge of the environment develops together with perception and becomes more detailed as the observer encounters more objects and events (253). According to the psychologist Allan Paivio, such tunings of the perceptual system might indeed underlie representations that are gleaned from past experiences (1990: 36).

When viewed in the light of modern experimental studies on perception, Gibson's proposal of perceptual knowing can be situated as an augmentation of, rather than a replacement for, the existing models of learning that are based on memory processes. This understanding is evident in Gibson's writing as well when he situates the perception of an environment as being contiguous with its conception ([1979] 1986: 258). The ecological approach addresses certain stages of our perceptual experience and complements higher-level mental processes. In that respect, Gibson's model of invariants aligns with the previously described cognitive models, such as perceptual symbols and schemas. This link is reinforced by the role of invariants in such models:

> Given repeated encounters with a set of objects that share certain features, birds for example, the neural units responding to the most invariant features (e.g., feathers and beaks) will grow into a highly interconnected functional unit, whereas the more variable features (e.g., color, size) will be excluded from the set of core elements. ... An assembly of neurons that forms in this fashion will exhibit many of the properties Barsalou attributes to perceptual symbols. It will be schematic, in that it represents only a subset of the features that any actual object manifests at any given time. It subserves categorization, in that the

same assembly responds to varying instances of some class of objects that have features in common. It is inherently perceptual, dynamic, and can participate in reflective thought. (Schwartz, Weaver, and Kaplan 1999: 632)

The complementary roles of perceptual and conceptual knowledge can be extrapolated to evolutionary processes on a grander scale. The physiological adaptation of an organism to its environment informs its behavior in terms of perceptual knowing. As Huron states, the *stable* features of an environment are instrumental in the formation of innate behavior over generations (2006: 61). On the other hand, behaviors that pertain to the rapidly changing aspects of an environment need to be learned. Although *instinctual knowledge* may appear to contrast the idea of prestructured perceptual knowledge, it should be noted that Gibson's model does not deny the role of the nervous system in registering the differences between invariants (1966: 284).

We engage with events in multiple stages of cognition. We gather structured information about events through their invariant features; the perception of these features helps us develop mental representations that aid cognitive processes on later encounters with similar events. Perceptual knowing can be associated with the evolutionary traits of the auditory system that deal with the low-level attributes of sound. For instance, the reverberant properties of an acoustic environment affords us a sense of spatial extent. The loudness and pitch of an environmental sound provide structured information about the nature of the associated event in terms of its spatial extent, proximity, and dynamic unfolding. In the following chapter, we will revisit the concept of affordances as they emerge virtually in the context of a musical narrative. Then, we will explore how such affordances relate to our sense of affect, which we discussed in Chapter 2.

Gestures in Electronic Music

As I mentioned in the previous section, the phenomena that we encapsulate into coherent structures, namely *events*, serve as the units with which we make sense of and describe what transpires around us. In this section, we will explore how this tendency to extract self-contained units from our experiences comes into effect when we listen to electronic music. To differentiate the events in electronic music as the product of a creative initiative that comes into play while not only composing but also listening, I will characterize them as gestures. As I will further elaborate in this section, gesture will function as the trace unit to which the poietic and esthesic processes apply. The concept of gesture is admittedly loaded with interpretations from various disciplines ranging from psychology to human–computer interaction. More importantly, it is extensively used in music research to discuss both the embodied and the conceptual aspects of musical movement. As I employ the results of the study to relate this concept to electronic music, these interpretations will help me not only align the current discussion with existing literature on musical gesture but also formulate a consolidated definition of gesture as a unitary structure in electronic music.

The common definition of gesture describes it as a movement of the body or an action that conveys an idea, meaning, or intention (*Oxford Dictionary of English*). Likewise, the composer Fernando Iazzetta characterizes gesture in music as a movement that can express and embody a special meaning (2000: 260). Movement, in this context, can be that of both the performer and the sound itself (Leman 2012: 6). While Gritten and King similarly define musical gesture as a movement or a change of state, they stipulate a need for intentionality for a movement to become a gesture (2006: xx). According to the music theorist Robert Hatten, gestures are both biologically and culturally grounded in communicative movement; through the intermodal interactions between our perceptual and motor systems, these movements are transformed into gestures that identify "significant events with unique expressive force" (2003). Hatten therefore defines *human gesture* as "any energetic shaping through time that may be interpreted as significant" (2006: 1). The significance of such a shaping lies in its ability to convey an affectively loaded communicative meaning (3). Our interpretation of a gesture relies on the cognition not merely of an object but of an event, whose movement displays a functional coherence with "the gestalt perception of *temporal continuity*" (2). Placing a similar emphasis on the act of cognition, the composer Wilson Coker broadly defines gesture as a recognizable formal unit that signifies musical or nonmusical objects, events, and actions (1972: 18).

These perspectives highlight some of the more conceptual underpinnings of gesture that extend beyond its interpretations related to embodied movement. Delalande offers a gestural taxonomy that draws an explicit distinction between the conceptual and corporeal uses of the term: With *effective gestures*, which are the mechanical movements necessary to activate an instrument, and *accompanying gestures*, which constitute the performer's bodily engagement that is not fundamental for sound production, Delalande (1988) refers to the embodied manifestations of gesture. In a third category, namely the *figurative gesture*, he highlights a metaphorical use of the term. A figurative gesture is perceived without a corresponding physical movement (Cadoz and Wanderley 2000). Iazzetta relates this third category to the musicologist Bernadette Zagonel's idea of *mental gestures*, which serve as models of experience stored in memory as products of inner hearing inherent to listening and composition processes (2000: 262).

The concept of musical gesture can function as both an analytical device and a compositional construct. In his analysis of gesture in contemporary music, the composer Edson Zampronha describes that the 1980s saw gesture-based composition overtake the parameter-driven techniques from the early years of electronic music. Zampronha (2005) criticizes these early techniques for their overreliance on "non-motivated combination[s] of parameters," echoing Xenakis's aforementioned criticism of total serialism as an irrationally complex technique for music composition. Zampronha characterizes the trend toward gesture-based techniques as an effort to "ground music in nature" by taking references and signification into account and by treating gestures as movements of sonic entities. Zampronha describes gesture as a musical unit of delimited configuration wherein parameters are treated interdependently. The interdependence of parameters leads to a gestalt perception, wherein gestures are viewed as "holistically perceived chunks" (Godøy 2006). Regardless of compositional technique or intent, the perceived interplay between the parametric configurations that make up a gesture can induce event gestalts for the listener. As Stockhausen describes, even though a composer might treat the individual attributes of sound, such as timbre, pitch, intensity, and duration, as separate properties, we perceive sound events as homogeneous phenomena rather than composites of these properties (1962: 40). Similarly, Hatten (2003) depicts musical gestures as "emergent gestalts that convey affective motion, emotion, and agency by fusing otherwise separate elements into continuities of shape and force."

In a study on the communication of expression in instrumental music, Vines et al. show that seeing the performer while listening to music impacts our emotional experience (2011: 157). But even in the absence of a performer, we mentally simulate physical gestures: the perception of instrumental music involves a *motormimetic component*, in which the listener mentally reenacts the articulatory gestures of a performer (Godøy 2006: 155). This motormimetic component relies on the listener's mental repertoire of action/gesture consequences (Leman 2012: 5). The physical actions involved in an instrumental music performance are therefore interwoven with how the resulting sounds would be perceived. This relationship, however, can disappear in electronic music since the direct corporeal link between the performer and the material is often broken. Producing a sound in the electronic medium is "rarely the result of a single, quasi-instrumental, real-time, physical gesture" (Smalley 1997: 109). A gesture in electronic music can encompass many sub-gestures performed at different times and layered together in a way that obscures the perception of the individual actions. By consolidating these various perspectives on gesture, we will try to arrive at a unified definition of this concept as it applies to electronic music. Paramount to this effort is the understanding that the way our cognitive faculties deal with a daily environment is not intrinsically different from how we navigate a piece of electronic music. Viewing gestures in electronic music as a counterpart to events in the environment, we can formulate the following definition:

A Gesture in Electronic Music …

… Is a Meaningful Narrative Unit …

When the human mind processes information, it looks for hierarchies and structural units to form systematic organizations (Özcan & Egmond 2007: 198). We utilize these meaningful units to navigate through the progression of our experiences. This is why Gibson considers events as the timescale of the environment ([1979] 1986: 12). Similarly, we can consider gesture as the scale at which a piece of electronic music moves forward. Gestures function as cognitive units by which listeners draw meaning from their experience as they segment an auditory stream into self-contained structural components.

In everyday life, we formulate narratives from these sensory streams by remembering the past and anticipating the future (Roads 2015: 323). In the following chapter, we will explore the concept of narrative in greater detail. For the time being, I will broadly define narrative as the story we impose upon a sequence of events demarcated by a beginning and an ending. Here is an example of an everyday narrative: "I was half asleep when my phone rang. I rushed to pick it up thinking it was you, but it turned out to be a sales call. I put down the phone and went back to my nap." Here, multiple events (rushing to pick up the phone, finding out who it was, putting down the phone) are encapsulated by a beginning (waking up from a sleep) and an end (going back to sleep). These events are tied together on a narrative arc through a sense of anticipation (i.e., waiting for a call from someone). By way of anticipating, we assign meaning to events (i.e., why did the phone ring?). Based on a similarly broad interpretation of narrative, Roads views the birth, development, and death of a composition as the integral parts of "a sonic narrative" (Roads 2015: 318). As we listen to a piece of music, our experiences accumulate into a narrative with an anticipated ending. We use this narrative to contextualize the individual gestures and imbue them with meaning.

As we touched upon in Chapter 2, Meyer groups musical meaning into two categories (1956: 35). A *designative meaning* is communicated when a stimulus indicates an event that is different from itself in kind (e.g., a word designating an object that itself is not a word). Conversely, an *embodied meaning* is established when the two are of the same kind. These categories can be applied to the semantic relationship between electronic music and environmental events as follows:

Electronic music —*embodied meaning*→ Environmental sound —*designative meaning*→ Event

Gestures in electronic music and everyday sounds are of the same modality. Through the principle of source bonding, the semantic relationship between the two is embodied. An environmental sound, on the other hand, communicates a designative meaning pertaining to an event, and it cannot exist as a disembodied phenomenon stripped of its causality. Although the composer and the listener meet in an absence of multimodal cues that could simplify the negotiation between a concept and its percept, sounds nevertheless evoke event-related information in multiple modalities (Warren, Kim, and Husney 1987: 326). The semantic relationship between an electronic music gesture and an environmental event can be severed by obfuscating the embodied meaning that links a gesture to an environmental sound. This, of course, is a principal technique of musique concrète.

I have previously referred to the experimental study by Dubois, Guastavino, and Raimbault, who found that most participants classified environmental sounds based on either source or action properties (2006: 867). In another experiment on environmental sound categorization, participants consistently used the source of a sound as a metonym to describe a *sound event* (Guastavino 2007: 55). A similar approach was apparent among the participants of the current study. For instance, in *Birdfish*, the same gesture was described as "water" and "boiling" by different participants. In another example, two participants used "fly by" and "bird wings" to denote another gesture. In these examples, action and object descriptors indicate the same gesture in an interchangeable manner.

In a study on the identification of environmental sounds, Marcell et al. (2000) produced a set of twenty-seven labels based on the descriptors provided by participants in a free association task. These labels were primarily based on object types, followed by event types and finally on the location or the context within which a sound is heard. Using a similar methodology, Gygi, Kidd, and Watson (2007) conducted another study on the categorization of environmental sounds. While the results show a consolidation of some of the labels generated by Marcell et al. into a total of twelve groups, both studies reflect a similar categorical distribution, which is partly mirrored in the current study, as seen in Figure 4.1. The primary difference in this case is the prominence of the conceptual and perceptual descriptor categories. This can be explained by the fact that the descriptors provided in the current study do not pertain to one-shot samples of environmental sounds and are instead evaluated within the broader context of an entire musical piece. Therefore, it is not surprising that semantic activations evoked by electronic music are more

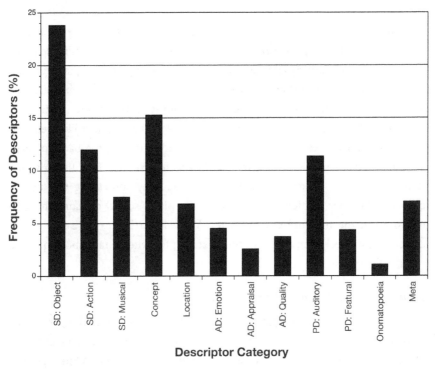

Figure 4.1 Overall categorical distribution of real-time descriptors.

conceptual than responses to samples of environmental sounds. Nevertheless, the source (SD) and location categories combined represent an overwhelming portion of the descriptors gathered from the study. The prominence of the perceptual descriptors category can be attributed to abstract musical gestures that do not occur in everyday situations; these kinds of descriptors are often utilized when source identification fails.

From a poietic standpoint, a gesture can be abstract and aimed at eliciting an emotion or a perceptual awareness. It can also be representational with the intention of triggering mental imagery. From an esthesic perspective, such intentions can be irrelevant since the listener will draw the meaning of a gesture, as with any object of perception, by way of situating it among their lived experiences (Nattiez 1990: 9). The semantic play between abstract and representational meanings can be highly subjective and, moreover, fluid. In *Diegese*, a particular moment in the piece was marked by separate participants with both source descriptors—such as "creature," "insects," and "bugs in my ear"—and affective descriptors—such as "creepy," "fear," and "weird." These indicate different forms of cognitive engagement with the same acoustic phenomenon.

The use of traditional musical forms and sound sources within electronic music reveals an interesting facet of musical meaning that can be exploited to evoke affective states. In *Christmas 2013*, instrumental sounds instill a sense of familiarity into a relatively alien sound world. Regarding the final piano gesture in *Christmas 2013*, a participant with no musical background submitted the descriptor "sounds like music." In their general impressions, another participant reported that although most of the sounds caused them to feel like being in "a place not on this earth," the piano sound made them "come back to earth and reminded [them] that it was music [they were] listening to." Similarly, one participant wrote, "in an imaginary world, suddenly something real begins to move." Another participant referred to the piano sound as "something to hold onto in the insecure environment." We will further explore this aspect of meaning in electronic music in the next chapter when I talk about quoting music within music.

... *Operates within Causal Networks* ...

As Gibson describes, ecological events can be nested within longer events ([1979] 1986: 110). We make sense of the passage of time through the coherent relationships between the events we observe in our daily lives. While a gesture is a temporal unfolding in and of itself, a multitude of gestures can mark the temporal unfolding of a higher-level form that represents a causal network. In Roads's dramaturgical model of sonic behavior, a gesture in electronic music can perform three activities: entering, acting, and exiting (2015: 324). In acting, a sound can stay the same, change in some way, become something else, or interact with other sounds. Such interactions reveal causal relationships between gestures:

> Interactions between different sounds suggest causalities, as if one sound spawned, triggered, crashed into, bonded with, or dissolved into another sound. Thus the introduction of every new sound contributes to the unfolding of a musical narrative. (Roads 2015: 328)

In literature, a *causal network* is a product of the readers' narrative interpretation of the text. These networks "represent the relationships between the causes and consequences of events in a story" (Gerrig and Egidi 2003: 44). Recalling our definition of narrative from earlier, the gestures in electronic music can be viewed as operating in similar networks. Causalities in electronic music can be abstract or concrete in nature. An abstract causality is the outcome of an arbitrary relationship defined conceptually or by convention. Between the 3'07" and 3'24" marks in *Element Yon*, a harmonic progression resolves to a pitch that is approximately 25 cents below B2. While the build-up toward this root is particularly apparent between 3'17" and 3'20", the resolving gesture is spatially distinct from this progression in an almost disembodied fashion. This gesture unfolds through an abstract causality by way of conforming to the listener's expectation of a harmonic resolution.

Concrete causalities, on the other hand, become apparent when a sound object imitates the behavior of a physical object. Causal relationships between sound objects, and particularly the adherence of abstract sound elements to a physical law, is a recurring technique in the original works discussed here and in my artistic practice in general. In Chapter 3, I referred to such causal systems as Sonic Rube Goldberg Machines. For instance, a grain repeating at gradually shortening intervals with its pitch shifting up as it nears a steady state can give the impression of a spherical object bouncing under gravity. It might then come in contact with another object and set it on a new motion trajectory. Imitating the chain reactions of a Rube Goldberg Machine, one can compose a choreography of sonic causalities. In such systems, the behavior of a gesture will affect those of concurrent and adjacent gestures. This concept was used to construct a number of gestures in *Diegese*: At 0'36", four consecutive transients build up to a final impact, behaving like toppling domino pieces that eventually collide with a bigger piece. At 0′48″, a reversed reverb tail builds up to a gesture that was often described by participants as bouncing balls. Finally, at 1′07″, a stretched piano sound glides upward in pitch and triggers a rapidly descending arpeggio, reminiscent of the toppling or bouncing pieces from before. These three examples, although using different sonic materials, share the same causality structure—a buildup that sets in motion a more substantial event.

As a composer constructs a sonic experience, they also establish a framework of expectations. Roads states that a sense of causality between sound events is necessary to achieve predictability: if listeners fail to associate these events with an underlying syntax, the piece turns into "an inscrutable cipher" (2015: 328).[1] We build causal networks from our narrative experiences. In literature, it has been shown that events that belong to such a network are much easier to recall than those that don't belong to it (Gerrig and Egidi 2003: 44). On a more visceral level, an event becomes easier to perceive when it conforms to expectations (Huron 2006: 43). That being said, any contrast between the predicted result of an event and its actual outcome amplifies the emotional response to it (22). The final section of *Element Yon*, which I have previously described as being composed to obfuscate any motivic closures to the piece, exploits the listener's reliance on causality. In the study, this resulted in several descriptors and impressions relating to anticipation. In their general impressions, one participant wrote, "Unpredictable. I liked that a lot." Later, in their real-time descriptors, this participant marked the final section of the piece with the word "unpredictable." At an adjacent moment in the piece, another participant submitted "when it's over, I can't tell" as a descriptor. Many other impressions relayed a similar sense of uncertainty. Accordingly, the narrative structures outlined by the participants who listened to *Element Yon* were noticeably more ambiguous than those provided in response to the other works.

In *Christmas 2013*, the juxtaposition of a Christmas carol with causally unfolding electronic gestures was intended to evoke a sense of nostalgia in a postapocalyptic environment devoid of human beings. An inexperienced listener wrote in their general impressions that while the opening of the piece felt familiar, the subsequent dissipation of the melodic material caused the piece

---

[1] While cautioning against the risk of stripping a piece of all predictability, Roads also acknowledges the power of strategically used juxtapositions (2015: 328).

to take a turn, which they would later refer to in their real-time descriptors as "suspenseful":

> It started to sound like bits and pieces of sounds and noises that I failed to make sense of. But these sounds, when they are together, they gave me this tense, mysterious feeling I don't know why. I feel like they would fit to a dramatic, tense moment of a film.

... Coexists with Other Gestures in Various Temporal and Spatial Configurations ...

A gesture in electronic music can range from the briefest sound that we can perceive to the longest sound that we can recognize as having a discrete form. This quality of gestures is also shared by environmental sounds: both the sound of an engine working throughout the day and the sound of an email notification going off once represent singular events. Regardless of their temporal extent, we manage to discern them as self-contained occurrences. Here, we are reminded of Lakoff and Johnson's comment on the human predisposition to impose boundaries on physical phenomena.

Unlike the gestures that can be performed with acoustic instruments, an electronic gesture can be decoupled from duration (Barrett 2012, personal communication). The electronic medium allows for a gesture to be extended beyond and contracted below the temporal extent of a physically performed musical gesture. Furthermore, the stretching or compressing of a gesture can be used to produce all-new gestures. In *Diegese*, some of the gestures that follow the causality model of a Rube Goldberg Machine are time-stretched versions of one another. While some of these operate on the same timescale with minor variations in temporal extent, a time-stretched and granulated version of this material acts as a ground gesture throughout the entire development section of the piece.

The elasticity of time afforded by the modern electronic medium creates many opportunities for building patterns. Gestures can be duplicated a virtually unlimited number of times. Using the rhythmic grid of a DAW, these duplications can be made perfectly periodic. Multiple gestures can also be aligned in time or in complex metric configurations. Through such temporal manipulations, existing gestures can be sawn into new gestures in the form of meso patterns. In *Touche pas*, between 2′26″ and 2′47″ a pulse is repeated at different rates in concurrent layers. The layer that exhibits the fastest rate is an almost continuous buzzing sound that amounts to a high-frequency texture. In a parallel layer, the pulses are spread apart enough to be perceived as discrete events but close enough to be considered parts of a larger whole. On account of the pitch variations between the individual pulses, the resulting gesture functions almost like a melodic line. Similar gestures at faster and slower rates create intermittent melodic phrases in different ranges. Finally, the layer with the slowest rate functions as a pedal point with pulses stretched into bass notes.

Just as we can distinguish between simultaneous events transpiring in our immediate surroundings, we can also make a meaningful segmentation of concurrent gestures in music. This is why Stockhausen describes multilayered spatiality as not only a composition technique but also a prevalent feature of human experience (1989: 106). This feature can be exploited in electronic music through the superimposition of layers as described above. The resulting temporal colocation of gestures creates opportunities for spatial articulations. In *Diegese*, between 0′24″ and 0′49″, gestures of different timescales are layered on top of each other. In the layer that is the farthest in terms of its spatial position, an ambient texture persists throughout the entire section. Moving closer, a low-frequency texture is initiated at 0′32″. Both of these layers are heavily reverberated to accentuate their distance to the listener. In a concurrent layer, another gesture pulsating at the granular scale establishes a third texture. Although this layer is in closer proximity to the listener than the first two layers, it is stripped of a figure role through the audio decorrelation of the left and right channels: an interchannel delay of 40 milliseconds widens this layer, pushing it to the margins of the stereo panorama. Lastly, in a fourth layer, another gesture consisting of transient elements assumes an unmistakable figure role as it traverses spatial extent that has already been established by the first three layers. Some of the descriptors submitted in this portion of the piece address the layers separately: While "ambient," "saw," and "sense of space" refer to the farther textural layers, "bugs in my ear" characterize the audio decorrelated layer since the stereo separation technique involves hard-panned copies of the same signal that evoke an inside-the-head sensation. Finally, "someone at the door" identifies the front and center qualities of the transient figure elements. Such descriptors demonstrate that the listeners can not only distinguish between coexisting gestures spatially and temporally but also prioritize ground elements at any given time.

... *Implies Intentionality*

Unlike environmental sounds, gestures are part of a communication system. Here, intentionality functions as a fundamental quality that separates a gesture in electronic music from an environmental sound.[2] Gritten and King argue that for a sound to be marked as a gesture, "it must be taken intentionally by an interpreter, who may or may not be involved in the actual sound production of a performance" (2006: xx). Gestures evoke mental imagery, which bears intentionality "in the sense of being *of*, *about*, or *directed at* something," whether that something is real or unreal (Thomas 2010).

Poietic actions of the electronic music composer result in intentional gestures. Even when an electronic gesture is composited from numerous actions performed separately, the result can embody a single intention. That being said, not all gestures are the outcome of a poietic initiative, for instance, when algorithmic processes are employed. Electronic music composition can indeed encompass approaches that are devoid of any poietic narrative arcs. However, the composer's conception of a musical work, in terms of its goals and techniques, will not always correspond to what is perceived by the listener (Smalley 1997: 107). For instance, a participant expressed in their general impressions that *Element Yon* is likely a generative piece, even though its sound material was the result of human performance. Conversely, some participants have associated some of the algorithmically generated gestures in *Diegese* with choreographed narratives.

Referring to the work of Jean Piaget, the acoustic communication researcher Ute Jekosch points out that gestalt forms are equally constructed by the observer as they are afforded by the observed (2005: 208). In other words, a musical signal reaching our consciousness is as much about us as it is about the music (Oram 1972: 56). The gestalt perception of a gesture can be rooted in either the actions of the composer or the listener's interpretation of these actions. This understanding coincides with Molino's communication schema, which situates *trace* as an embodied artifact at a neutral level that is devoid of a communicational hierarchy between the producer and the receiver. The act of esthesis performed by the receiver, in this case the listener, is imposed upon this trace, but the traces of the poietic process left in the symbolic form are not always perceived by the receiver:

> The esthesic process and the poietic process do not necessarily correspond. ... [T]he listener will project configurations upon the work that do not always

---

[2] Environmental sounds, when used in the context of music, can function as intentional gestures.

coincide with the poietic process, and do not necessarily correspond to what Deliège happily dubbed "realized intentions." (Nattiez 1990: 17)

In other words, the sender and the receiver do not have to come to the same understanding of the trace; this viewpoint supports Marcel Duchamp's idea that the artist and the viewer represent two poles of an artistic experience, where an agreement between the artist's intention and the viewer's interpretation is not imperative (Frisk and Östersjö 2006: 8). The casting of the neutral trace into a gesture is a function of intentionality: by imposing a unitary function to the trace, the listener extracts an intentional gesture from the physical artifact, but conflicts of intentionality between the poietic and esthesic processes are impossible to avoid. Esthesic intentionality will result in gestural hierarchies that may or may not serve the narrative goals of the composer, yet these will nevertheless adhere to the listener's construction of a story. In the study, there were numerous instances where the participants inferred different narrative intentions from their experiences than what I had planned for (or what I recognize) in these pieces. In other cases, they caught on to my plan even when I considered the creative intention behind a gesture to be relatively obscure. Besides such parities and disparities, a transposition of intentionality between the composer and the listener is also possible. Here is a general impression response from the preliminary study conducted with *Birdfish*:

The sounds heard and experienced by a baby in its mother's womb prior to birth, and its eventual coming to earth.

While this impression does not reflect the specific story that I was trying to communicate with this piece, it does bear an uncanny resemblance with my narrative plan in terms of its metaphors and story arc, which involves organic forms that evolve as they travel from beneath the ocean into the sky. Once the participant constructed their own narrative, my poietic intentions seem to have been mapped on to that narrative. Gestures that give a sense of a cavernous, underwater environment were interpreted as the inside of a womb. Having established such a setting, the participant contextualized the remainder of the gestures accordingly.

Based on the qualities that we discussed in this section, we can arrive at an idiomatic definition of the electronic gesture as an intentional narrative unit that coexists with other gestures in various temporal and spatial configurations within causal networks. Most of the properties that factor into this definition are tightly interwoven: The coexistence of gestures is articulated through the causal network that they operate in, which in turn is a function of their intentionality, and vice versa. While sound-producing events in our environments also bear meaning, coexist in various temporal and spatial configurations, and operate within causal networks, gestures are set apart as narrative units in electronic music on account of their intentionality. Relying on its poietic and esthesic dimensions detailed in this section, the electronic gesture can be utilized as an analytical tool, particularly when dealing with the narrative structure of a work. To that end, the next chapter will explore how such structures materialize within the broader context of a piece. Adopting a worldmaking approach to the listening experience of electronic music, I will introduce the concept of diegesis as it relates to this experience. I will provide examples from the study to highlight the relationship between the gestural qualities presented here and the listener's construction of a narrative from the totality of a piece.

5

Worldmaking in Electronic Music

In our discussion of meaning in electronic music in Chapter 2, I situated listening as a fundamentally creative act performed by both the composer and the listener. Then in the previous chapter, where I formulated a definition of the electronic gesture as a structural unit that is comparable to an event in the environment, I argued that the intentionality of a gesture can stem from not only the composer's poietic actions but also the listener's construction of meaning from the trace. In this section, I will expand upon this view by framing the act of listening as one of *worldmaking*. In discussing the listener's engagement with electronic music through this lens, I will borrow the narratological concept of diegesis and discuss how various interpretations of this concept can help us portray the spatiotemporal realities that listeners infer from works of electronic music. I will juxtapose some of these interpretations to establish a narrative framework for electronic music informed by a range of artistic disciplines, including literature, film, theater, and visual arts. We will explore the narrative function of gestures by looking at how the momentary impressions that a gesture prompts are situated within the broader context of a piece. Later in the chapter, I will revisit the theory of affordances, this time as it relates to our experience of affect, which we covered in Chapter 2. Formulating a correlation between the two, I will introduce the concept of diegetic affordances that stem from the implied universe of a piece. In the final section of the chapter, I will discuss the idea of musical quotations in the context of electronic music, looking at the ways in which a piece of music can be made a diegetic element in another work and how such quotations inform the listener's interpretation of this work.

Diegesis

In his book *Ways of Worldmaking*, the philosopher Nelson Goodman argues that any creative process, whether it's the painting of a picture or the formulation of a scientific theory, entails the creation of a world version that stems from the frame of reference that we use in the relevant domain. We employ "words, numerals, pictures, sounds, or other symbols of any kind in any medium" to make these world versions (Goodman 1978: 94). However, we can only test these versions by comparing them to worlds that are already described, depicted, or perceived. In other words, worldmaking starts from worlds already on hand—"the making is remaking" (6).

Goodman identifies five ways of worldmaking: composition and decomposition, weighting, ordering, deletion and supplementation, and deformation. *Composition and decomposition* are common ways of worldmaking that are often used in conjunction; we take things apart and build them back up in a new way. While two worlds may consist of the same elements, by *weighting* these elements differently, we can cluster them into separate kinds with distinct emphases. Even when these elements are weighted similarly, their *ordering* can reveal new periodicities and proximities. Making a world from an existing one can also be achieved by either *deleting from* or *supplementing* the world at hand; deletion, in this context, can merely imply overlooking or reducing. Finally, we can reshape an existing world into a new one by way of *deforming* it. Not only can these processes be reframed to yield different results, but they can also be combined to achieve alternative ways of worldmaking (ibid.: 17).

Although Goodman offers this taxonomy as a cross-disciplinary framework, it maps onto musical creativity without extensive reinterpretation: We compose musical phrases, emphasize certain musical qualities for articulation, supplement and deform phrases to create variations, and order these in different ways to build forms. In musique concrète, we deform and order sound recordings to compose world versions that are free from familiar objects—versions that point inward rather than outward. In algorithmic music, we compose systems that are capable of composing world versions, wherein the algorithm or the generative procedure underlying the system replaces the active participation of the composer. Regardless of the creative method, every piece of music establishes a world version by being a sample of that world (Kulenkampff 1981: 257). Even abstract works that are devoid of any representation can inform our worlds by way of symbolizing their internal features through exemplification (e.g., of patterns and formal properties) or expression (e.g., of movement) (Goodman 1978: 105). Rejecting the dichotomy between emotive and cognitive processes on the basis that emotions function cognitively, Goodman argues that aesthetic experiences are intrinsically cognitive; we gauge and grasp artworks and integrate them with the rest of our experience of the world (Goodman 1968: 248). This way, our encounters with art afford new world versions, "new structures of appearance and of reality" (Elgin 2001: 687).

Recalling our discussion about the threads of communication in electronic music from Chapter 2, we can characterize poietic and esthesic acts intrinsically as worldmaking. The composer's act of worldmaking in electronic music can be direct, delegated (e.g., to a process, algorithm, or system), or involve a combination of both. A corresponding, if not as self-evident, creative act of worldmaking takes place during the esthesic process. In the previous chapter, I emphasized meaningfulness and intentionality as core properties that position gestures in electronic music as counterparts of events in the environment. I stressed that these properties do not need to originate from the poietic process and can be rooted in the esthesic thread between the material and the listener.

Here, I will adopt the narratological concept of *diegesis* to situate the listener's act of worldmaking in the broader experience of electronic music. The narratologist Gérard Genette defines diegesis as the spatiotemporal universe that a narrative refers to (1969: 211). Although I will offer different perspectives on this concept, Genette's interpretation will serve as the operative definition of diegesis throughout this chapter and the next. As I will detail shortly, Genette builds a narrative framework around this concept as it applies to literary fiction, articulating the relationship between the author, the reader, and the text. However, the scope of this interpretation can be expanded to encompass other forms of artistic expression, including electronic music. To this end, I will leverage a worldmaking approach to diegesis and propose that every piece of electronic music, regardless of its production technique or the perceived representationality of its sonic material, can evoke in the listener's mind an implied universe.

An Interdisciplinary Contextualization of Diegesis

The concept of diegesis can be traced all the way back to Plato, who categorized poetic modes into mimesis (i.e., imitation) and diegesis (i.e., narration). This concept has since been adapted to many artistic disciplines as a means to describe narrative constructs in art and to situate the agents of an artistic phenomenon in relation to one another. Although there are numerous interpretations of diegesis that are seemingly divergent, these interpretations commonly outline domain-specific relationships between the artist, the artwork, and the audience. In Platonic mimesis, past, current, or future events are presented through imitation, whereas diegesis involves the act of narration (Plato [c. 380 BC] 1985: 247). Therefore, in Plato's taxonomy of narrative forms, theater is mimetic, because actors imitate (i.e., reenact) situations, while poetry is diegetic because the poet, speaking in their own voice, recounts an event or situation as a narrator who is external to the world of the story.

Needless to say, artistic forms and the ways in which they are presented to audiences have evolved substantially since Plato's time. For instance, poetry back then was not an art form that readers engaged in on their own. Instead, poets recited their works to audiences in public gatherings (Nagy 1990: 21). Since then, the transformations in how we create, distribute, and consume art have prompted new and domain-specific interpretations of diegesis. In film, for instance, the sounds that originate from the characters or objects in a scene (e.g., a dialogue between two characters, or music coming from a radio in the scene) are considered diegetic. On the other hand, the film score that is external to the universe of the story (i.e., inaudible to the characters in the film) is considered non-diegetic sound. Therefore, whether a film sound is diegetic or not is based on whether it is internal or external to the world of the story that is being shown on the screen.

Diegesis also serves broader functions in film theory. A prominent perspective in film studies postulates that every film is diegetic because the director decides which parts of a story's universe will be displayed on the screen. Even when the world shown to the audience is a perfect replica of the world outside, the director's framing of this world is considered an act of narration. And when the director assumes the role of a narrator, all that transpires on the screen becomes illusory (Hayward 2006). But we could then question which form of art escapes this criterion. Even a generative work of art necessitates a moment in time when the artist initiates an algorithm, thereby establishing a narrative context for the piece framed by the progression of time. From this relativistic point of view, all artistic material could be deemed to display diegetic features through the artist's framing of a world version even when this material is presented in a representational medium that is mimetic in the Platonic sense.

Genette applies the concept of diegesis exclusively to literary theory as a device for situating the author, the reader, and the elements of the story (e.g., characters, venues, time) in relation to one another. Delineating the nested layers of a narrative by starting from the physical world of the author on the outermost level, Genette outlines the concept of diegesis as the spatiotemporal universe to which the author's narration refers. Therefore, in his terminology, a diegetic element is one which "relates, or belongs, to the story" (translated in Bunia 2010: 681). Here, we observe a thread emerge between Genette's literary definition of the term and its aforementioned use in film, where a diegetic sound originates from a source that belongs to a scene.

Coalescence of Mimesis and Diegesis

In Plato's dichotomization of mimesis and diegesis, the primary difference between the narrative actions pertinent to each mode, namely reenactment versus recounting, is whether the medium of representation remains the same as that of the represented or, in other words, whether there is a mediation between the act of expression and what is being expressed. While a dialogue between two characters remains a verbal exchange in theater, it would be delivered in the medium of text in literature. Electronic music, in this sense, is mimetic. While it may represent extramusical events, it does so through sonic connotations. It is not narrated like diegetic poetry but speaks for itself; it is representational not in the form of a mediator but as a portion, or an abstraction, of reality—*a world version* that exemplifies internal or external auditory qualities. The loudspeaker will detach the sound from its source, but the medium of the phenomenon remains unchanged. Here, we can revisit the semantic link between electronic music and environmental sounds that we discussed in the previous chapter:

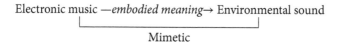

The embodied meaning that constitutes a semantic relationship between electronic music and environmental sounds is mimetic in essence because the medium for both is sound. However, electronic music does evoke more than memories of sounds, just as an environmental sound signifies more than the physical entity it emanates from. What is auditorily perceived can be the by-product of an event, but it is not the event itself:

> When we say that information is conveyed by light, or by sound, odor, or mechanical energy, we do not mean that the source is literally conveyed as a copy or replica. The sound of a bell is not the bell and the odor of cheese is not the cheese. ... Nevertheless, in all these cases a property of the stimulus is univocally related to a property of the object by virtue of physical laws. (Gibson 1966: 187)

Here, Gibson underlines the causal link between a stimulus and the object or phenomenon it originates from: the stimulus delivers *information about* the phenomenon that it specifies (ibid.). But every sound we hear triggers a semiotic web that prompts us to extrapolate from what the sound immediately represents. The sound of someone knocking on a door may immediately signify a hand impacting the surface of a door, but it also triggers a whole network of anticipation as to who is knocking on the door and why. A sound as simple as a knock on the door can therefore become evocative of affect. Regardless of how abstract a sound may be, once we associate it with a cause, it *unmutes* a context around itself.

The mimetic acting in a theatrical work prompts a similar reaction. What we witness on stage is just a portion of the world that we imagine and situate the characters within. Here, a bond between Plato's mimesis and Genette's diegesis materializes. Although a narrative form might be purely mimetic, it will nevertheless imply a spatiotemporal space different from the one that the audience inhabits. Electronic music *presents* to listeners sounds that *represent* events; it does not speculate about—or recount—sounds. Electronic music is therefore mimetic in the physical domain of the concert hall, but it creates a diegesis for the listener in the semantic domain. A diegetic link therefore emerges between the electronic gesture and the event that it may imply for the listener.

Presentationality

If a piece of electronic music is capable of casting an implied universe, where is the listener situated in relation to it? In the next chapter, we will explore some of the ways in which the listener can be given first-person or third-person views into a diegesis, or even be transported from one view to the other. But as a precursor to this discussion, I will once more turn to dramaturgy. In theater, *representational acting* is the kind of performance where the actor ignores the presence of an audience and leaves the viewers outside the universe of the story that the play depicts. This is unlike *presentational acting*, which not only acknowledges the audience but also addresses it. In certain theatrical forms, such as those that are improvisational or interactive, the engagement between the presentational actor and the audience can even alter the progression of the story.

The concept of presentationality finds use in other art forms as well. For instance, the visual artist Sanford Wurmfeld characterizes *presentational art* as a kind of art that is structured as a statement for an active viewer to experience. The sensory nature of this art form renders it untranslatable; "the ideas or feelings transmitted by it are tied to the particular object that expresses them" (Wurmfeld 1993). The artist Mel Alexenberg (2004) corroborates this perspective when he describes presentational art as a real thing that is presented as itself: "form and color presented as form and color."

Wurmfeld's characterization of presentational art as evocative of an untranslatable sensory experience reminds us of the Lévi-Strauss's portrayal of music as an untranslatable language. Indeed, Alexenberg likens presentational art to music, arguing that the two share an abstract presence that does not represent anything other than itself. While Alexenberg's framing of music lacks the distinction between embodied and designative meanings in music, it also highlights the presentational qualities of musical sound, which can manifest in electronic and instrumental music alike. In instrumental music, the presence of an external performer *presenting* a composer's work becomes an explicit aspect of the listener's experience. The performance is not merely a representation of the work but the embodiment of the work itself. As I will discuss in the next chapter, presentational qualities in electronic music can materialize in two ways: through the composer's presence in the work, or through the listeners' awareness of their physical selves.

Narrativity

The philosophers Gilles Deleuze and Félix Guattari argue that the greatest challenge an artist faces is to make an artwork stand up on its own. Overcoming this challenge requires "great geometrical improbability, physical imperfection, and organic abnormality" from the perspective of lived perceptions and affections

(2000: 465). This is reminiscent of Goodman's proposal that worldmaking starts from worlds already at hand. In Chapter 3, we discussed the idea of *the abstract as a negation of reality*: an improbability is intrinsically defined through our preconceived notions of what is probable; we can only tell if something is abnormal by drawing from our experiences of normality. In the arts, the tension between the two manifests itself as a negotiation between the audience and the artwork. For instance, a literary text that demands too much interpretation prompts the reader to naturalize it by using acquired knowledge and resolve semantic inconsistencies by turning to its narrative structure (Mikkonen 2011: 113). In doing so, an assumption of *world semantics* is transferred from the real world to the fictional world (Bunia 2010: 699). This inherent reliance of abstraction on reality is what renders impossible fiction "an ostensible oxymoron" (Ashline 1995: 215).

A narration can invite the audience into a vast universe by providing only a limited amount of information about the events and characters in this universe (Bunia 2010). The diegetic details that an artwork inevitably leaves out prompts the audience to fill in the gaps by building a semantic network within and beyond what is presented to them. This license on the audience's part is further apparent in Souriau's interpretation of diegesis, which he describes as "all that belongs, 'by inference,' to the narrated story" (Gorbman 1980: 195). This inference inherent to the esthesic process lies at the core of the viewer's act of worldmaking. Although there is no narrator in nonvocal music comparable to that in a literary text, listeners partly assume the role of the narrator by constructing stories out of their experience of the narrative.

A similar perspective on artistic experiences is reflected in *the principle of minimal departure* (Ryan 1980). This principle posits that we structure our interpretation of alternative realities as closely as possible to our own reality by projecting things we know about the real world on the implied reality of a story (406). This way, the possible worlds of fiction lie in actual worlds; unmaking and remaking familiar worlds will yield sometimes obscure but ultimately recognizable worlds (Goodman 1978: 104). A narrative can therefore restrict the information it communicates about a universe to a small set of the actors and events that populate it (Bunia 2010: 686). But by virtue of the principle of minimal departure, we are able to form "reasonably comprehensive representations" of such worlds even though they are always described incompletely (Ryan 1980: 406). Here, one of the ways of worldmaking that Goodman proposes, namely *supplementing*, comes into effect. As much as we can overlook missing aspects of a world version, we can also supplement it to make it coherent. This is how an implied spatiotemporal universe can be logically complete even when it's insufficiently narrated.

Genette describes narrative as a type of discourse that signifies a story (1980: 27). Similarly, the cultural theorist Mieke Bal defines narrative as a medium in which a story is told. Bal further unpacks this definition by characterizing story as "a fabula presented in a certain manner" (1997: 5). Here, Bal splits what Genette refers to as a story into two layers, namely fabula and story. A fabula, which consists of a series of chronologically connected events, can be framed into different stories in ways that will prompt different inferences. The reader's interpretation of the text is an extraction of the fabula, which will inevitably be manipulated by the story (9). In this context, ordering and weighting, two of Goodman's ways of worldmaking, become useful narrative devices. In their construction of the story, the author, for instance, can expand and contract the pace of certain events in the fabula, or change the sequence in which they are presented, to give them different emphases. Bal stresses that a text can be understood as a narrative in any medium including "language imagery, sound, buildings, or a combination thereof" (5), suggesting that her model is applicable to other artistic modalities. If we were to situate Bal's taxonomy in Molino's semiotic model presented in Chapter 2, *text* would correspond to the trace, whereas *fabula* would be the outcome of the receiver's esthesic interpretation of this text as influenced by the producer's story:

Producer —*poiesis* (story)→ Trace (narrative text) ←*esthesis* (fabula)— Receiver

Adapting Bal's taxonomy to musical narrative, Meelberg maps text to perceivable sounds, story to musical structure, and fabula to a series of musical events that the listener stores in long-term memory (2006: 82). Meelberg describes musical narrative as a "representation of temporal development," adding that while the narrative signifies a succession of events in a process, it is not the process itself (39). The narratologist David Herman defines narrative more broadly as "a basic human strategy for coming to terms with time, process, and change" (2009: 2). Here, we are reminded of Roads's interpretation of narrative as a story that the human brain constructs by relating the present to the past and anticipating the future. From this viewpoint, it could be argued that it is inevitable for listeners to extract narrative structures from their musical experiences due to the simple fact that a piece of music encapsulates an event or a series of events between a starting point in the past and an anticipated ending in the future.

In instrumental music, the narrative emerges in the abstract realm of musical sound, structured around musical expectations that are grounded in both cultural and perceptual traits of music. Through such traits, music can narrate forms that unfold over time (Meelberg 2006: 1), which can then translate to emotional experiences (Walton 1994: 60). As we discussed in Chapter 2, electronic music adds to this experience a distinct layer of representationality made possible by the electronic medium; the transition from music as a physical phenomenon to music as it is appraised by the listener is intercepted by a layer of meaning attribution. Listeners inhabiting the physical domain of the concert hall superimpose mental representations over their embodied experience of the sounds. The affective impact of the musical work is immediately informed by this act. These representations are rooted in our own life experiences, expectations, and convictions. As a result, they can diverge and overlap from one listener to the other. In Chapter 3, I described that *Birdfish* narrates a story of creatures striving to transcend the surface of the ocean as they morph from underwater beings into avian creatures. Here are excerpts from two general impressions that this piece elicited:

> Participant A: I heard robotic bugs moving around being commanded by more intelligent robotic beings. There was water, stepping into water, robotic dialogues and also progress made by the robotic bugs in their task.

> Participant B: This music reminded me of a cartoon I used to watch when I was in high school. I related the piece to the story of the cartoon which told the struggles of liquid-like alien creatures who on the one hand were not from this world but on the other hand had to adapt to survive.

The real-time descriptors submitted by Participant A consist of narrative elements similar to those that are found in their general impressions. They mention "bugs" as the workers who are "flying" and "walking" while "making progress" on a "project" that they are working on collectively. The real-time descriptors submitted by Participant B substitutes the aliens mentioned in their general impressions with other creatures, such as "baby bird," "huge ant," "snail," and "worm." Despite this change of characters, a similar story persists with such real-time descriptors as "sent to earth," "can't fit in," and "struggle again." The timing of the latter descriptor coincides with those of

Table 5.1 Comparison of Characters, Settings, and Actions Identified by Two Participants Who Listened to *Birdfish*

	Character	**Setting**	**Action**
Participant A	(Robotic) bugs	A project	Making progress
Participant B	(Alien) creatures	Earth	Struggle

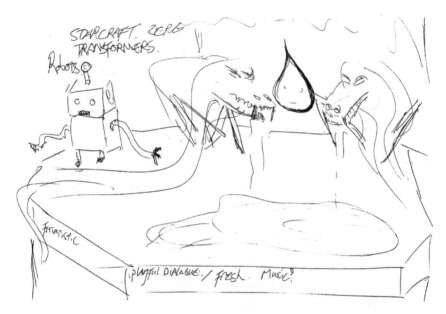

Figure 5.1 A participant's general impression of *Birdfish* in the form of a drawing.

Participant A's descriptors such as "some adjustment" and "project continues." A correspondence chart between these two accounts can therefore be constructed as seen in Table 5.1.

In these cases, we can observe two seemingly distinct fabulae constructed from the same trace. However, there is an apparent pairing between how these two implied universes are populated and how the actors inhabiting these universes behave. Besides the metaphorical similarities between the actors (i.e., robotic bugs vs. alien creatures), the unfolding of the events that relate these actors to one another follow similar narrative arcs. A third participant's interpretation of the same work takes the form of a drawing, as seen in Figure 5.1. In their real-time descriptors, this participant makes the same references to StarCraft, Zergs, and robots as they did in their drawing. Zerg is one of the species in the popular strategy video game StarCraft. Displaying insect- or reptile-like properties, members of this species work in swarms to form a civilization. Here, we observe the emergence of concepts, actors, and actions similar to those reported by Participants A and B.

Diegetic Affordances and Affect

In Chapter 2, we touched upon Massumi's definition of affect as a pre-personal intensity that signifies the transition of the body from one experiential state to another. This perspective was rooted in Spinoza's interpretation of affect as a change in the body's capacity to act. Massumi expands on this view by framing affect as the body's potential for interaction, which can then be qualified into semantically formed progressions (2002: 35). Let's compare this interpretation of affect with Gibson's theory of affordances. As we discussed in the previous chapter, Gibson describes affordances as action possibilities that an environment provides to an observer; the information carried by an affordance is structured before it reaches the observer, but the interpretation of this information as an action possibility will be relative to the perceiving body.

The theory of affordances and the concept of affect are rooted in two separate fields of study, namely psychology and philosophy. But, following the central premise of cognitive science, an inquiry into the synergy between the two can yield new insights. Table 5.2 displays how the core properties of affordances compare to those of affect.

Here, we can observe a phenomenological continuity between the two, one being a property of an object and the other being an affection of the body that encounters the said object. Both are unqualified capacities, one representing a potential that the object presents to the perceiver and the other representing a

Table 5.2 Correspondence between the Underlying Properties of Affordances and Affects

Affordance	Affect
Pre-personal, structured information available in the (material) environment	Pre-personal intensity
Precedes cognitive processes	Unqualified experience preceding semantic processing
Action possibility	Potential for interaction
Relative to the observer's form	A corporeal phenomenon

potential of the body to react. In their discussion of affect, Deleuze and Guattari suggest a progression from material to sensation in aesthetic experiences, wherein "the plane of the material ascends irresistibly and invades the plane of composition of the sensations themselves to the point of being part of them or indiscernible from them" (2000: 466). In his seminar *Continuous Variation*, Deleuze even describes the relativity of affection to the perceiving body, pointing out that his perception of an object will be different from that of another organism (1978: 6). Forging a link between these two concepts, we can characterize affordances as evocative of affects. The affordances of the environment provide the body with capacities to transition from one state to another.

In Chapter 2, I briefly spoke of Massumi's portrayal of emotion as a personal capture of affect—that is, an intensity qualified into meaning. Despite his characterization of emotion as a sociolinguistic fixing of the experiential quality that is affect, Massumi deemphasizes the unidirectional succession of affects into emotions by promoting the recursiveness of affect as a "simultaneous participation of the virtual in the actual and the actual in the virtual, as one arises from and returns to the other" (2002: 35). Higher mental functions, such as cognition and volition, "are fed back into the realm of intensity and recursive causality" (30). Affects, anchored in physical reality, can therefore be both pre- and post-personal. This dualistic approach to affect is also hinted at in Freud's positioning of unconscious affects in immediate adjacency to conscious thoughts, practically inseparable from cognition (Seigworth and Gregg 2010: 2).

Gibson offers a similar perspective on affordances rooted in virtuality when he describes how our apprehension of visual media, such as film and photography, relies on two kinds of awareness: a direct awareness gleaned from the physical properties of the medium, and an indirect awareness of the virtual objects and environments presented in this medium (Gibson 1986: 283). These objects and environments, while virtual, can also stimulate affective experiences. When watching a movie, the audience is aware that they are in a viewing environment separate from the implied universe of the film. But once they are acculturated into the story of the film, a mundane and seemingly nonaffective act, such as a character turning on the lights in a dark room, can become loaded with affect. Gibson describes how film evokes in the viewer an intense level of empathy in the form of an awareness of being in the place and situation shown on the screen. That being said, the viewer cannot act upon this situation; what they experience is a passive form of visual self-awareness and visual kinesthesis (Gibson 1986: 295). Along a similar line, electronic music can arouse a passive aural kinesthesis. For instance, one of the participants who listened to *Diegese* offered the following general impression:

> Glass/metal ping pong balls are constantly being dropped on the floor as we walk through an empty salon with bare feet; we leave this room and go out in a jungle, moving through the grass stealthily; passing through cascading rooms; we arrive in another salon.

While many of the objects employed in this narrative also appear in other participants' descriptors, details like "walking with bare feet" and "moving stealthily" are indicative of this participant's individual affective engagement with the diegetic environments that they visualized. A diegetic element represented in electronic sound suggests a material of second order: a musical gesture, on its surface, is an informationally structured object that the mind can extract a representation from (Nussbaum 2007: 23). Alongside the "irresistible ascent" of embodied sound to affects that are grounded in our direct awareness of the material, the implied material of the representation ignites an affective thread of its own. The spatiotemporal universe evoked by this representation will have its own dimensions, landscapes, surfaces, and objects. As the listener observes the implied possibilities that this universe affords, they also experience the affects associated with them. I will refer to such possibilities as *diegetic affordances*.

The study results show many instances where diegetic affordances prompted higher-level semantic associations, which ultimately informed the listener's act of worldmaking. Such affordances are often linked to features of electronic music that are corporeally relevant to the listener. Reverberation, for instance, affords a relative sense of space, which can be further manipulated by adjusting the frequency response of this space. In *Birdfish*, a reverberation with a medium-length tail paired with low-frequency rumbles was utilized to establish the sense of an expansive yet enclosed environment. This elicited such descriptors as "cave," "dungeon," and "big spaceship." While the kind of environment may be dictated by the listener's narrative interpretation of the piece, the features of this environment are informed by the spatial affordances of the sound. Similar spatial and spectral cues in *Christmas 2013* prompted listeners to submit "open sea," "open space," and "sky" as descriptors. For some listeners, such cues implied an affordance of locomotion, which in several cases manifested itself as "flying."

Diegetic affordances can cue us about not only environments but also objects placed within those environments. In *Element Yon*, the frequency and amplitude modulations in certain gestures instigated such descriptors as "metal balls getting bigger and smaller" and "high tone falls and hits the ground." In these examples, distinctly perceptual qualities are situated in metaphors while retaining their embodied relativity to the listener. A similar example is observed in the responses to some of the gestures with high-frequency content in *Birdfish*. The participants characterized these gestures with descriptors like "ice," "glass," "metal," "blade," and "knife." These descriptors imply both a metaphorical association and an affordance structure between high frequencies and a perceived sense of sharpness.

Indeed, we regularly extract information about the make of an object from the sounds it makes. As the design researcher William Gaver describes, the material, size, and shape of a physical object will govern how the object vibrates and therefore produces sounds: for instance, vibrations in wood damp much more quickly than they do in metal, "which is why wood 'thunks' and metal 'rings' and big objects tend to make lower sounds than small ones" (Gaver 1993a: 7). In electronic music, the affordance structure between an object and the elementary attributes of the sounds that it makes can be exploited to suggest physical causalities. This is an area where granular synthesis bears a distinct potential. Beyond the metaphorical relationship between a sound grain and a particle, the granulation of a sound can emulate the behavior of a physical process that an object is going through. By manipulating the envelope, rate, density, and pitch properties of the granulation, different kinds of materials can be made to undergo various physical transformations.[1] In the study, gestures produced through granular synthesis were often characterized as particles (pieces, cells) of various materials (glass, metal) that are dividing (breaking, splitting) into pieces and coming back together (merging, colliding).

Researchers have previously demonstrated the relationship between the fundamental frequency of a sound and the perceived size of the object that this sound indicates (Coward and Stevens 2004). The frequency and amplitude envelopes of a grain can be altered to specify the size of a perceived particle. *Touche pas*, for instance, has prompted many real-time descriptors that refer to spherical objects of diverse proportions. Figure 5.2 shows a drawing submitted as a general impression of *Touche pas*. Furthermore, the timbral characteristics of grains can be altered in order to imply different surface materials. In *Diegese*, which quotes a particular granular texture from *Touche pas*, listeners

---

[1] For a more extensive analysis of the relationship between granular synthesis and physical modeling, see Roads's *Microsound* (2001: 97).

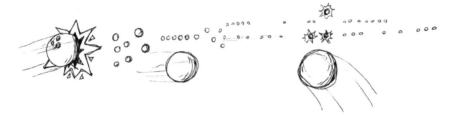

Figure 5.2 Another general impression expressed in drawing, showing the "bowling balls," "ping pong balls," and "mostly circular shapes" of different sizes that a participant imagined while listening to *Touche pas*.

differentiated between timbral varieties by defining various material types and objects. Separate participants described "glass/metal balls," "ping pong balls," a "pinball machine," "champagne" (cork sound), a "woodpecker," and "knocking on the door." Here the materials vary from metal to plastic to wood. *Touche pas* evoked a similar spectrum of materials as evidenced by such descriptors as "coins," "marbles," "ping pong balls," "bowling ball," and "xylophone."

The perceived physical affordances of a gesture produced with granular synthesis are intrinsically informed by the evolution of the individual grains or grain streams over time. The grain behavior employed in *Diegese* is inspired by the concurrent loops of uneven durations heard in Subotnick's seminal piece *Touch*, a behavior that Roads also utilized in *Touche pas*. When multiple loops are blended together, the resulting texture implied for most participants a sense of physical objects "bouncing" (e.g., "marbles bouncing") or "falling" (e.g., "rocks falling together"). One participant wrote, "the clicking sounds … resembled a dropped ball bouncing on a surface, since each sound came in slightly quicker than the previous one." Another participant described *Touche pas* as displaying a "convincing physicality." Once the motion trajectory of a gesture is coupled with the perceived material of the object that the gesture evokes, higher-level semantic associations occur: One participant characterized the physical behavior of grains in *Touche pas* as "bouncing on wood," which prompted this participant to imagine a "marimba." Another participant who listened to the same piece imagined "marbles" instead, which made them think about "childhood," "fun," and "games."

Our tendency to associate the acoustic properties of a sound with the featural aspects of its source extends beyond our perception of inanimate objects. The linguist John Ohala (1983) highlights a cross-species association of high-pitch vocalizations with small creatures, and low-pitch vocalizations with large ones. The size of an animal, as implied by the fundamental frequency of its vocalizations, indicates to other animals whether it poses a threat. Many descriptors submitted by the participants allude to liveness or living organisms. These descriptors were often prepended by featural qualifiers to form such noun phrases as "tiny organisms," "baby bird," "little furry animal," "huge ant," and "huge animal." Here, featural descriptors signify the proportions of the perceived organisms in relation to the listener. In these cases, featural information extracted from the sounds afforded the listeners a spatial hierarchy between the imagined creatures and themselves. In the next chapter, we will further explore the affordance of liveness in sound as it relates to our semantic interpretation of electronic music.

Music as a Diegetic Actor

The layer of meaning attribution that comes into effect while listening to electronic music has an interesting effect on listeners' engagement with more traditional forms of musical material in the context of electronic music. Such forms could include a tonal melody, a discernible rhythm, or even a gesture that displays a timbral similarity to an acoustic instrument. While these forms would be expected to cause an affective appraisal in an instrumental music context, the study results revealed a meta-evaluation of such forms when encountered in an electronic music piece. Prior to an affective appraisal, the listener identifies the phenomenon as the musical form that it is, situated in the universe of the piece. That is to say, abstract musical elements essentially become concrete, functioning as diegetic objects in the context of the piece in the style of *a television in a movie scene*.

As I remarked in Chapter 3, one of the gestures in the representational sound world of *Birdfish* is an abstract leitmotif consisting of three notes in octaves played legato. In certain instances, this gesture is supplemented by subtle resolutions in the lower spectrum that remain in a territory between underwater rumblings and a sub-frequency pedal point. The piece is otherwise almost devoid of any material that could be aligned with tonal structures. Among all the narrative components of the piece, which are conspicuously set underwater, this leitmotif creates a moment of stark contrast. A participant of the preliminary studies described the final recurrence of this leitmotif at the very end of the piece as a "musical climax." Out of all the descriptors submitted by this participant, such as "water dripping off of cave walls" and "factory noises," this is the first and only reference to a musical form. It is also interesting that the leitmotif inspired a need to pronounce the *musicality* of a gesture in a piece of music. An audience member from a performance of *Birdfish* characterized this leitmotif as a "musical souvenir." This expression appropriately illustrates the diegetic quality assumed by a familiar musical form in the context of an electronic music piece.

Quoting Music within Music

Quotation has a long history in music. Charles Ives famously quoted other composers' works from different styles to create semantic connotations in his work (Ballantine 1979: 168). The third movement of Luciano Berio's *Sinfonia* is a collage of pieces from a number of composers including Debussy, Ravel, and Stravinsky, among others. The history of musical quotations extends far beyond the twentieth century. For instance, many Baroque composers quoted Protestant chorale tunes in their works for organ and vocals (Holm-Hudson 1997: 17).

Given its inherent capability to capture and reproduce sounds, the electronic medium grants the composer an "unprecedented ability to include exact quotations from pre-existing sources" (Beaudoin 2007: 149). Early examples of such quotations are found in Vladimir Ussachevsky's *Wireless Fantasy*, which cites Richard Wagner's *Parsifal*; Pierre Henry's *Comme une symphonie envoi a Jules Verne*, which appropriates Anton Bruckner's symphonies; and Karlheinz Stockhausen's *Hymnen*, which is composed of recordings of various national anthems from around the world. In his 1985 essay, the composer John Oswald introduced the concept of plunderphonics as both a subgenre of electronic music based on the sampling of existing musical tracks and a statement on issues of copyright and ownership as they relate to recorded media. His 1989 album *Plunderphonic* recycled material from a wide range of artists including Michael Jackson and the Beatles; the undistributed copies of the album were destroyed due to claims of copyright infringement (Collins 2010: 66).

Two of the pieces used in the study explored this idea of quoting music within music. As I described in Chapter 3, *Diegese* consists of two quotations: the first is a recreation of the granular texture from *Touche pas*, and the second is a snippet of a recording of Beethoven's Opus 90. Since the former is an emulation of a texture rather than an exact quotation, its effect appears in the study results only in the form of semantic similarities between the descriptors submitted for *Diegese* and *Touche pas*. These similarities were extensively interpreted above. However, in the latter quotation, the listeners could clearly discern the piano segment even if they were not able to point out a particular piece as a source. Two-thirds of the participants made note of the piano sound in their real-time descriptors. One participant included an appraisal descriptor and referred to the quotation with "nice piano" followed by "lovely" at 1′15″. Another participant made note of the quotation in their general impressions as "the frantic piano sound."

In *Christmas 2013*, a similar quotation is taken from the Christmas carol *Silent Night* as it is played by a jazz trio. While the entirety of the piece was composed out of processed extracts from this recording, between 0′4″ and 0′20″ the quotation can be clearly made out. Furthermore, brief references to the quotation are sparsely distributed throughout the remainder of the piece. As a result, musical source qualifiers made up the most salient descriptor category for *Christmas 2013* with 19 percent of all descriptors submitted for this piece. This is also the highest ratio of musical source descriptors among all five pieces.

Three-quarters of all the participants who listened to *Christmas 2013* used the real-time descriptor "piano" at least once. Participants who did not directly refer to the piano alternatively submitted "music I know," "bar, dance," and "ballet," indicating an engagement with the meta-musical qualities of the quotation. Subjects who participated in the preliminary study that gathered general impressions of *Christmas 2013* provided concurring feedback. The quotation was characterized with such impressions as "real instruments," "acoustic instrument," "a phantom harmonium," and "cozy trio." Unsurprisingly, "Christmas" was another salient descriptor. The final ambient crescendo was described by two participants as being reminiscent of "Pink Floyd."[2] Besides descriptors that denote musical instruments or forms, participants also submitted quality descriptors such as "familiar" and "cliché." A music technology student who participated in the study offered an intriguing perspective on the structuring of the piece by defining "planes in a dimension." They described the impact sounds, which articulate the temporal and spatial unfolding of the piece, as establishing a plane. According to this participant, "rooms" represent another plane. More interestingly, they refer to the quotation also as a "cliché" and explain that this too becomes a plane on its own. I find this *bracketing* of the cliché to be an apt

---

[2] In hindsight, I now hear a brief moment in this segment that can be reminiscent of the intro to "Shine on You Crazy Diamond" from the Pink Floyd album *Wish You Were Here*, which was formative in my musical taste growing up. Although this was not an intentional quotation, the comparison holds sentimental value for me.

description of my poietic intent, which was to assign a diegetic function to the quotation.

As I described in Chapter 3, my goal with contrasting electronic and sterile sounds of *Christmas 2013* with those that are lo-fi and organic was to create a sense of *future nostalgia*. One participant described their experience as a "memory of an event" and associated their sense of "something recalled from the back of the mind" to the tonal thread of the piece. One of the participants denoted that, besides feelings of "flight" and "movement," "nostalgia and space" dominated their experience with the open ending with the piano enhancing their feeling of "nostalgia/longing." Another participant wrote:

> I had the impression of being in the air (like an angel) and moving over a city on Christmas evening. The sounds escaped from human festive activities, some became distorted and merged over with other ones, others still referred to remembrances (nostalgia) of childhood (acoustic instruments).

As we observed in the categorical distribution and the correspondence analysis in Chapter 3, *Christmas 2013* yielded the highest ratio of not only musical source descriptors but also affective quality descriptors across all five pieces. Two contrasting groups of affective quality descriptors emerged in the responses to this piece: one group consisted of such descriptors as "smooth," "mellow," "familiar," "childish," and "relaxing," while the other included "incongruent," "creepy," "dark," and "weird." These descriptors reveal a duality that was expressed through various means in most of the general impressions (e.g., earth/nostalgia vs. space, imaginary vs. real, childhood vs. distance/melancholia). This result can be interpreted as the outcome of a possible priming caused by the musical quotation early on in the piece. The precedent this opening sets with a relaxing, mellow, and familiar feeling amplifies the disorienting sensation of the ensuing diegesis.

As I mentioned in Chapter 4, in response to the final piano part in the piece, a participant with no musical background wrote the descriptor "sounds like music." Another participant mentioned in their general impressions that although most of the sounds caused them to feel as if they were in "a place not on this earth," the piano sound made them "come back to earth and reminded [them] that it was music [they were] listening to." Along a similar line, one participant wrote, "in an imaginary world, suddenly something real begins to move." Another participant with no musical background referred to the quotation as "something to hold onto in the insecure environment."

Although *Diegese* also incorporates a clearly discernible piano recording, the categorical distribution for this piece shows a markedly lower ratio of musical source descriptors than that of *Christmas 2013*. This difference is likely due to the extent and form of the quotations. In *Christmas 2013*, the listeners could not only recognize a multiplicity of instruments but also identify a musical form in the quotation. With *Diegese*, while two-thirds of the participants used "piano" as a descriptor, the only other descriptors used to denote the quotation were "jazz" and "melody."

A Diegetic Actor as Music: Electronic Music and Science Fiction

The artistic choices made by film and game sound composers to incorporate electronic sounds in order to evoke various concepts and emotions in the audience has a peculiar effect on the experience of electronic music, especially among the listeners who are less experienced with this type of music. With *Birdfish*, science fiction was a prominent point of reference for the participants as evidenced in such descriptors as "Star Wars," "R2D2," "Starcraft," "spaceship," and "robot." Various general impressions for other pieces also delineated similar concepts: "I get images of science fiction: spaceships etc.," "[synthetic sounds] reminded me of a world that you may find in a movie like Tron" (*Element Yon*); "sounded like a soundtrack for a horror or science fiction movie" (*Diegese*); "astronomical documentary, museum of science" (*Christmas 2013*). Some participants associated their experience more loosely with movie soundtracks in general: "It reminded me of sound effects used for tense moments in thrillers," "I feel like they would fit to a dramatic tense moment of a film" (*Christmas 2013*). In these cases, memories of diegetic film sounds acted as a reference for the listener's musical experience. Two participants offered more specific descriptions of this relationship: "[I imagined] sound design people working on a sci-fi film, enjoying their work," "the brief tonal sounds reminded me of R2D2" (*Birdfish*). The latter is a fairly reasonable account of how a sound as generic as a sine wave can be connotative of a robot. The sound of R2-D2 utilizes gliding pitch variations similar to those that are indicative of intonation in speech. However, due to listeners' tendency to maintain a semantic coherence across various gestures, unintended references to diegetic sounds of science fiction might predispose the entire story, as evidenced in some of the examples provided earlier in this chapter. In these cases, film or game sounds that are diegetic to the universe of the story serve as external referents that ultimately influence the listener's interpretation of the work.

As we briefly touched upon in Chapter 1, the inherent role of technology in electronic music, and the unfamiliarity of general audiences with this kind of music in its early days, prompted many film composers to exploit electronic sound for dramatic effect in science fiction films. Here are two comments on this trend that are roughly three decades apart:

> The imaginative faculty of listeners to electronic music is largely influenced by its utilization in the mass media, above all as background music for science fiction programmes, a fact that is also naturally connected with the properties of the material. ... [T]he entire sound material of electronic music today already meets a perception that is semantically prepared in this direction and has therefore already undergone a social conventionalization of its meaning. (Karbusický 1969: 36)

> [The] sense of disorientation produced in some listeners by the impact of electronic sounds was the basis of the early use of electronic sound-materials for science fiction productions. The inability of the listener to locate the landscape of sounds provided the disorientation and sense of strangeness which the producer wished to instill in the listener. (Truax 1996: 139)

Seeing that the current study was carried out approximately two decades after Truax's remark, the perceived coupling between electronic music and science fiction has yet to wane. Throughout the twentieth century, certain electronic sounds have become culturally codified as having scientific and futuristic connotations. Regardless of how far in the past this codification may have occurred, we still associate these sounds with objects of science fiction. Especially sound designs based on simple waveforms passed through rectangular envelopes, colloquially referred to as bleeps and bloops, maintain a strong association with robots and computer systems. Today, we commonly encounter such sounds as notification alarms in household appliances and traffic management systems. Yet the word "alarm" was submitted only on two occasions in the entire study, whereas there were at least one or two references to sci-fi concepts in response to each piece. It is up to the composer whether to avoid or exploit this affordance of electronic sound, but science fiction remains to be a prominent source of imagery for electronic music listeners.

6

Tracing the Continuum

In this chapter, I will compare some of the perceptual aspects of the electronic music experience with those that are predominantly semantic. I will first group these under two seemingly separate domains of experience in electronic music, namely the physical and the semantic domains. We will then examine how these two domains can come in contact with each other while we listen to a piece of music. As I parse through the explicit and implicit features of these domains, I will discuss the ways in which the listener can be situated inside or outside the implied reality of a piece. Rather than merely contrasting these features, I will highlight the role of their coalescence in actively shaping our experience of electronic music. In the final section of this chapter, I will offer a case study, where I analyze Natasha Barrett's piece *Little Animals* through the lens of the theoretical constructs presented here.

Domains of Experience

In Chapter 2, we explored some of the biological and cultural factors that influence our appreciation of music. We also looked at neuroscientific research on mental mechanisms that underlie this process: Juslin and Västfjäll underscored how these mechanisms are not exclusive to music appreciation and are in fact the same as those that we rely on in our daily lives. Existing studies on auditory cognition and how the human mind deals with everyday acoustic stimuli can therefore help us understand listeners' reactions to electronic music. Throughout this section, I will introduce findings from both the current study and other empirical research to trace out compositional strategies that draw on the perceptual and conceptual traits of the electronic music experience. I will tie these findings in with the theoretical perspectives presented in the earlier chapters to delineate how listeners' narratives are spun across the physical and the semantic domains.

The Physical Domain

The physical domain of an electronic music experience represents the empirical reality that the listener inhabits. Sound is a part of this reality in the form of a measurable acoustic phenomenon that interacts with the auditory faculties of the listener. Let's look at some of the ways in which the perceptual qualities of sound can shape the listener's interpretation of a work. Most of these qualities inform our perception of music regardless of the musical material or language. In the case of electronic music, these perceptual qualities will have an impact on semantic processing, which we will explore in the following section.

Awareness of the Physical Self

In the results I presented in Chapter 3, we observed that *Birdfish* had prompted a higher number of real-time descriptors than *Element Yon* with 5.96 descriptors per minute per participant versus 3.77. We also saw that the most salient descriptor category for *Birdfish* was *object source descriptors* (SD: Object), accounting for 33.83 percent of all descriptors submitted for this piece. When combined with *action* (SD: Action) and *musical source* (SD: Musical) descriptors, the representational sound world of *Birdfish* appears to have instigated a dominance of source identifiers accounting for 55.96 percent of all descriptors. With *Element Yon*, *concept descriptors*, which represent 28.64 percent of the real-time entries for this piece, were the most common. Given the abstract disposition of this piece, this result conforms to the expectation that the listeners would favor conceptual descriptors over source descriptors in their feedback. Furthermore, *Element Yon* produced a notably higher percentage of *affective descriptors* (AD), with *emotion descriptors* (AD: Emotion) being the most salient subcategory (18.75 percent for *Element Yon* vs. 6.95 percent for *Birdfish*). Whereas the spontaneous responses to *Birdfish* related more to descriptions of objects and actions, the listeners of *Element Yon*, besides using fewer descriptors to express themselves, were more inclined to reflect about their affective experience.

As we discussed in Chapter 2, failing to identify a source for a sound can prompt us to pay attention to its acoustic properties. This effect becomes particularly apparent when we compare the percentages of *auditory descriptors* (PD: Auditory) between *Element Yon* (17.18 percent) and *Birdfish* (8.45 percent). With the former, the listeners were more inclined to use descriptors that pertain to the spectral and dynamic properties of sound, supporting the understanding that abstract conceptualizations grounded in physical attributes of sound may emerge in the absence of source identification. The general impressions of *Birdfish* commonly took the form of word and sentence lists enumerating imagined sound sources and visually oriented narratives that recount a story involving such sources. On the other hand, general impressions of *Element Yon* were mostly concerned with concepts, such as "contrast," "flow," "hollowness," and "heaviness"; affective appraisals, such as "exciting," "curious," "calm," and "annoyed"; and the physical attributes of sound; even the impressions relating to objects or environments, such as "big magnets," "gravity," "circus-like," and "a dark metro station," were more conceptual than indicative of sound sources. One participant expressed that "no images came to [their] mind" while listening to this piece.

When the embodied experience of an electronic music piece disposes the listener to concentrate on perceptual qualities, the listener can become attentive to not only the physical attributes of sound but also their own physical presence. Given the abstract nature of *Element Yon*, which leverages a broad range of contrasting frequencies, it is not surprising that the listeners were inclined to point out the spectral properties of the piece in their descriptors. But more interestingly, the results have revealed a much more articulated sense of self in response to this piece when compared to the other pieces. The participants who listened to *Element Yon* commonly wrote their general impressions in first-person (i.e., "I felt …"). Conversely, the participants who listened to *Birdfish*, which consists primarily of representational sounds, commonly assumed the role of an outside viewer who observes and reports the unfolding of various events (i.e., "… happened"). This change in voice suggests that the prominence of representational elements in a piece can influence where the listeners situate themselves in relation to a piece. With *Element Yon* the listeners tended to place themselves at the center of the experience, whereas the listeners of *Birdfish* assumed the role of a spectator watching a narrative unfold. In that sense, the experience of *Birdfish* can be likened to that of watching a work of representational acting, during which the audience is situated outside of the diegesis. On the other hand, the self-awareness apparent among the listeners of *Element Yon* is akin to what an audience member experiences when they are called out by the actor during a play. In this case, music becomes a *presentational object* that addresses the audience by making them acknowledge their own embodied presence.

Additionally, spectral extremes at high amplitudes, such as very high and very low frequencies that are clearly audible, made the listeners conscious of not just themselves but also their act of listening. Almost half of the participants who listened to *Element Yon* expressed a form of annoyance with the high frequencies with such descriptors as "disturbing," "annoyed," or "irritating." Moreover, three-quarters of the participants particularly noted the rests in the piece either by pointing out the pockets of silence themselves or by describing the relief they evoked. Although *Birdfish* incorporates the momentary use of high-frequency gestures comparable to those in *Element Yon*, only one participant used the descriptor "harsh high" to indicate a similar annoyance. Furthermore, despite the similarities between how the silences are structured in each piece, none of the silences in *Birdfish* were denoted as bringing relief. This suggests a difference in engagement with perceptual qualities based on the representational capacity of a piece. In an abstract work where the listener is left to focus on the perceptual qualities of the work, the same spectral extremes tended to draw an amplified recognition of one's act of listening. This can serve as a compositional strategy for the composer to ground the listener's experience in physical reality by making the listener aware of their physical selves and their act of listening. As we will discuss later in the chapter, this also creates opportunities for transforming where the listener is situated in relation to the narrative layers of a piece.

Stream Segregation

Frequency and loudness are among the most salient features of musical sound. Modulating these properties over time can elicit visceral reactions that I have characterized in Chapter 2 as affect. This is because our auditory system has an evolutionary disposition to detect changes in pitch, dynamics, and tempo. These features therefore play a central role not only in musical articulation but also in our interpretation of auditory events in general. According to Moore and Hedwig, we can perceive a sequential stream of auditory events as coming from either a single source (i.e., as a product of *fusion*) or multiple sources (i.e., resulting from *fission*) (2012: 919). They argue that the perceptual differences between the individual elements of a stream dictate the extent to which stream segregation occurs (2002: 320). For instance, when the degree of separation between two sounds is high enough, a fission occurs. Differences of intermediate size bring about a state of *bistability*, wherein the percept flips between one or multiple streams (Moore and Hedwig 2012: 919). Individual or combined variations in spatialization, frequency, temporal envelope, and phase spectrum contribute to a separation between auditory streams.

Such variations are often used in music composition to establish figure and ground relationships between concurrent sounds. In an orchestral piece, a solo instrument might play a melodic line in forte while the rest of the orchestra accompanies this line in mezzo piano, leaving room for the dynamic and melodic articulations of the solo instrument (i.e., the figure). Here the figure and ground elements establish two streams that we distinguish from each other. In the visual domain, the figure and ground relationships are often established between concurrent elements. On the other hand, the auditory stream segregation model implies that figure and ground relationships in sound can be either concurrent or adjacent. In granular synthesis, which relies on streams of microsounds, the manipulation of the spectral and dynamic qualities of individual grains is often used to articulate textures that are analogous to a surface with dips and peaks. Using these kinds of articulations, multiple gestural streams can be created from a single series of grains, as exemplified in the opening section of *Birdfish*. By modulating the grain amplitude, grain length, and the center frequency of per-grain filters, two concurrent layers from a single stream of grains are established in the first 38 seconds of the piece. This was reflected in the listener responses with two separate yet concurrent themes where descriptors such as water and underwater were intermixed with creatures, animals, and birds. In *Diegese*, between 0′16″ and 0′25″, an arpeggio starts out as a traditional musical structure, but as the individual notes of the arpeggio are shortened, they begin to amalgamate into a percussive texture, which segues into the following section where two versions of the same texture breaks off into independent streams: while one of these versions assumes a more textural role, a sloweddown and spatialized version takes on the role of a figure gesture. Here, the temporal extent and pitch of the individual components of the gesture, as well as its spatial animation, set it apart as a foreground element in a dedicated stream.

Habituation

Numerous studies have investigated how changes in acoustic parameters contribute to the evocation of affect in music. For example, in an fMRI-based study of emotional and neural responses to music, Chapin et al. show that tempo fluctuations cause emotional arousal among both experienced (i.e., musically trained) and inexperienced listeners (2010: 7). In another experimental study on the perception of emotional expression in musical performance, Bhatara et al. conclude that timing variations alone have more impact than variations in amplitude, although the latter is also shown to communicate a significant amount of expressive information (2011: 932). Huron similarly draws attention to the impact of dynamic and temporal properties of sound on the affective appraisal of music. For instance, the frisson response is found to be correlated with loud passages in music (Huron 2006: 34). Focusing on such interactions between dynamic and temporal organization, Huron offers an intriguing perspective on auditory repetition. In what he refers to as an *orienting response*, individuals turn their heads in the direction of an unexpected sound as a basic reflex. However, when such a stimulus is repeated, the individuals will habituate to it after a while and cease the orientation reflex. A sufficiently novel change in the sound, however, will cause *dishabituation*, prompting the individuals to reorient to the stimulus (49). In the context of acoustic ecology, Truax describes *habituation syndrome* as an individual's adaptation to annoying noises in their acoustic environments as a result of extended exposure (1984: 90). He further characterizes habituation as a form of auditory desensitization. When a noise lacks an immediate significance, we are able to habituate to it and disregard it as background noise.

The affective impact of a musical phrase is often a function of its transient properties. As I described in Chapter 2, *Christmas 2013* begins with a sudden and loud gesture that is reminiscent of an orchestral stab. Indeed, most participants made note of this gesture, referring to it with such descriptors as "bam," "sudden start," and "percussion." While this staccato gesture recurs throughout the piece in different shapes and forms—sometimes quieter, time-stretched, or pitch-shifted—it also recurs in its original (i.e., loud and abrupt) form two more times. However, having heard this gesture at the beginning of the piece and its various incarnations thereafter, the participants left these later iterations of the gesture unmarked. Similarly, other impact sounds of comparable amplitude appearing throughout the piece were usually noted only once by most participants. Habituation, in this case, may have reduced the dramatic impact of these gestures over time. Furthermore, this implies that certain sounds that stand out perceptually can be made to be less protuberant through repetition even when they maintain their acoustic prominence.

Previous research has shown that unstructured musical sequences impose a much heavier memory load on the listener (Deutsch 1980: 8). In that respect, a clear rhythmic grid can be utilized as a device of relief in an electronic music piece that exhibits a temporal complexity matching that of everyday sounds. An example of this is found in *Touche pas*. Toward the end of this piece, which consists primarily of sparse and arhythmical distributions of granular sounds, the composer introduces a looping of a 3-second segment. This particular gesture stands out because the lack of such repetitive behavior up until this point makes it sound as if the piece is stuck in the final groove of a broken record. Indeed, once participant described this moment with the word "malfunction." Many other participants referred to this particular moment both in their general impressions and in their real-time descriptors. While one participant characterized it as being funny, another participant counted the repetitions, signaling that keeping track of this unexpected pattern was part of their musical appraisal. Other participants used such descriptors as "repetition," "counting," "rhythmical," "coda," and "loop," which indicate a distinct engagement with this unique occurrence of a repetitive structure in the piece.

Although individual events in daily life might exhibit periodic behaviors (e.g., an engine running), multiple periodic events almost never align on the same temporal grid. In other words, environmental sounds do not share a common tempo in the way that the individual parts of a musical piece might. This is why rhythmic patterns in sound can often be indicative of human-made systems, including musical forms. As a result, a rhythm in the context of an electronic music piece that lacks a rhythmic grid constitutes a powerful tool on account of its ability to disrupt the diegetic experience by evoking an immediate awareness of a musical structure. Amid the uneven temporal configuration of the gestures in *Christmas 2013*, a half-measure sampling of a drum beat at 1'26" was enough to elicit tempo-related descriptors such as "rhythm," "rhythm again," and "little beat," indicating a tendency to latch onto rhythmic patterns.

Based on *Element Yon*'s reliance on abstract sounds that reveal temporal and spectral contrasts, we could use Deutsch's (1980) terminology to characterize this piece as consisting of "unstructured musical sequences." The impact of this quality is apparent in the real-time descriptors, such as "exhausting," "chaotic," and "confused." Furthermore, in their general impressions, one participant stated that "sounds without a rhythm made [them] curious but at the same time they were really exhausting." A moment in the piece that this participant marked as being "exhausting" was marked by another participant as "I repeat and repeat but you don't get it." This participant had already elaborated on this section in their general impressions when they likened their experience to witnessing a redundant argument between people. Here, the lack of temporal or spectral patterns that could prompt habituation seems to have created a sense of futility in the listener's mind. This may be an undesirable musical sensation for some listeners: in their general impressions, one participant described that unclear sound structures made them feel bored.

The Semantic Domain

The semantic domain of an electronic music experience is the outcome of the listener's construction of meaning from what they hear. The implied spatiotemporal universe of a piece (i.e., the diegesis) emerges in this domain. Every new element in the piece primes the listener for what is to come: the flow of gestures instigates a constant semantic realignment, where the diegesis established up to a certain point in the piece constitutes an evolving semantic context for new information. This way, even when an explicit sound element is removed from the piece, it can persist implicitly. In the listener's mind, past diegetic actors interact with present ones as the listener fills in semantic gaps in ways that conform to the narrative that they glean from the work. By imagining the implied universe of a piece, the listener, for example, can extract semantic polyphonies from perceptually monophonic phrases by constructing implicit figure and ground relationships between diegetic actors. This act of worldmaking lies at the root of an immersive listening experience.

Effects of Semantic Context

In an experimental study on the effects of context when identifying everyday sounds, Ballas and Mullins found that contextual inconsistencies had a negative impact on identification (1991: 199). Similar effects have also been observed not only in the visual domain between images of objects and actions (Vigliocco et. al. 2002: B61) but also cross-modally between words and sounds (Orgs et al. 2006: 267). In another study by Ballas, identification time for a sound has been found to be highly correlated with and inversely proportional to the sound's causal uncertainty (1993: 250).

According to Wishart, contextual cues can influence not only our recognition of an auditory image but also how we interpret the events we hear (1996: 152). In *Element Yon*, the abstract nature of the sound material prompted the listeners to identify contextual cues with *concept* and *auditory descriptors* such as "surrounded," "panning," "stereo," and "reverb." In *Birdfish*, which utilizes similar spatial techniques, the contextual cues were often annotated with *location descriptors*. The auditory references to water and creatures, which were the two most persistent themes in the real-time descriptors for this piece, caused listeners to imagine possible environments, such as "underwater," "lake," and "aquarium." When the recognition of amphibian-like sounds was evaluated within a space articulated by reverberation, such descriptors as "cave" and "dungeon" became salient. Similarly, a combination of water-like sounds with the inference of a cave prompted such general impressions as "water dripping off of a cave wall" and "slimy rocks and stalactites." These kinds of conceptual combinations instigate higher-level associations besides what the individual components of such combinations could suggest alone. In other words, the semantic relationship between the actors can imply environments or even new actors. The coherence between an object and its context can activate semantic processes wherein the listener embellishes the diegesis with other appropriate objects that are not explicitly presented. As the neuroscientist Moshe Bar describes, recognizing an object that is highly associated with a certain context makes it easier for us to recognize other objects within the same context even though those objects can be ambiguous when viewed in isolation:

> Each context (for example, an airport or a zoo) is a prototype that has infinite possible exemplars (specific scenes). In these prototypical contexts, certain elements are present with certain likelihoods, and the spatial relations among these elements adhere to typical configurations. Visual objects are contextually related if they tend to co-occur in our environment, and a scene is contextually coherent if it contains items that tend to appear together in similar configurations. (Bar 2004: 617)

Corroborating the studies mentioned at the beginning of this section, Bar states that when an object is not congruent with its context, it is processed more slowly: items and relations that are typical to a context will lead to faster processing times than those that are atypical (618). In other words, a context gathered from the recognition of objects feed back into how the individual objects in the extracted context will be perceived. Bar proposes the concept of *context frames*, which are "prototypical representations of unique contexts" that are derived from real-world scenes (618). This concept is reminiscent of models of mental representation, such as *perceptual symbols* and *schemas*, that we discussed in Chapter 4. Prototypical contexts can indicate both the likelihood of an element to appear in a given context and its spatial configuration (617).

Once listeners establish a semantic context, they tend to hold on to it for the remainder of the piece. With *Birdfish*, the listeners who constructed a fabula from their experience have continually made references to descriptors that they had previously used or to their general impressions. Moreover, some of these recalls were not merely in response to the reappearance of a gesture but served to advance the overarching narrative. For instance, a listener of *Birdfish* who established a nautical scene early on in the piece with such descriptors as "underwater," "sand," "water," and "waves" described the ending of the piece with "big waves," "sea is projected in the air," "and explodes." Another participant, who constructed a story of robotic bugs working on a project, as we discussed in the previous chapter, described the ending of the piece with "workers are pleased," "big cheers," and "project successful." Here, we can see that both participants felt a need to address the climactic ending of the piece. But how this climax was situated in their respective fabulae shows a semantic coherence with the individual contexts that had already been established.

As we discussed in Chapter 2, the contextual segmentation of everyday sounds is dependent on those distinctions between sound events that denote sources and ambient noises in which background sounds are blurred together (Guastavino 2007: 57). Furthermore, the ambient noise context affects how a listener reacts to an environmental sound (Raimbault and Dubois 2005: 342). These everyday reflexes bring to mind our perception of figure and ground configurations in music, but their impact becomes even more prominent in the context of electronic music, in which everyday sounds themselves constitute a frame of reference. While the aforementioned blurring effect is a cognitive artifact of selective attention, the composer can impose this selection deliberately by obscuring sound events, and therefore building this cognitive artifact into the work. One such strategy is followed in the second movement of *Birdfish*: the organic gestures that assume figure roles in the first movement are subjected to low-pass filtering, reverberation, and frequency shifting. Although the temporal density of these gestures remains the same in the second movement, they are pushed back in the spatial structure of the piece and situated as a stage for new figure elements. These gestures, which now serve a ground function, blur together in the background not just perceptually but also on account of the functional contrast between the foreground and background layers. This is similar to how we pay analytical attention to short-term details in foreground listening, while we group concurrent background elements in gestalt patterns (Truax 1996: 59).

Semantic Gestalts

Meyer suggests that the connotative capacity of a phrase in instrumental music is intrinsically connected to how much it diverges from a "neutral state":

> A tempo may be neither fast nor slow; a sound maybe neither loud nor soft; a pitch may seem neither high nor low, relative either to over-all range or the range of a particular instrument or voice. From the standpoint of connotation these are neutral states. Connotation becomes specified only if some of the elements of sound diverge from such neutral states. (Meyer 1956: 263)

A ground element in music, such as an accompaniment texture, can be viewed as a neutral state from which a figure element, such as a melody, diverges from. Even when a melody and its accompaniment are played with a single instrument, we can distinguish these elements as separate parts based on the pitch, time, and dynamic differences between them. In everyday listening, our auditory systems perform a similar segmentation by listening to the foreground and the background simultaneously. In the previous section, we talked about the segregation of auditory streams based on their perceptual qualities. A similar segregation can also occur based on the information delivered through such streams. Through selective auditory attention, we manage to home in on certain sounds based on their semantic content even when they are not perceptually prominent. The semantic gestalts that we derive from auditory streams can inform how we compartmentalize gestures into figure and ground layers.

Spatialization, loudness, and filtering determine the perceived physical proximity of a gesture. In tandem, these parameters help establish a semantic gestalt of *motion*. Sounds that move between stationary speakers follow choreographies designed by the composer and imply for the listener an animation of objects in the absence of any actual moving sound sources. Regardless of whether these objects are abstract or concrete, the listener hears—and furthermore imagines— further beyond the mere changes in physical parameters and extracts a gestalt sense of movement emerging from the interplay between these parameters. Through the use of panning, reverberation, and filtering, a ground texture can be made to be recognized as a *location*. This way, the texture transcends its background musical function and suggests a metaphor of place. Once a location is semantically associated with this texture, a scene is established for successive figure gestures, which will then be evaluated by the listener in reference to *where* they occur. This conditioning can operate both ways, as the semantic content of a figure gesture will inevitably feed back into how a consecutive ground texture is interpreted. The spatial movement of foreground gestures can establish a sense of location, for instance, when the "spatial activity of insects" conjures up images of a "hive" or a "swarm."

Signs of Life

Guastavino's aforementioned experiment on environmental sound categorization reveals that the presence of human activity is one of the most salient features for the hedonic judgment of urban soundscapes (2007: 54). Moreover, soundscapes that predominantly feature human sounds are perceived to be more pleasant than those that involve mechanical sounds (61). In a similar vein, adopting the semantic criteria for acoustic ecology proposed by Murray Schafer and Bernard Delage, Raimbault and Dubois construct a soundscape classification consisting of "road traffic (car—truck—motorcycle), other transportation (railway, aircraft), working machines (street cleaning, working site), music, people's presence (speech, walking), and nature (wind, animals)," emphasizing the prominent role of human sounds in our evaluation of our acoustic environments (2005: 343). As a species, we are sensitive to sounds of human origin. For instance, self-representation sounds, which are associated with bodily functions such as breathing and heartbeat, have been found to increase the corporeal awareness of the listener with a stronger capacity to induce emotional experiences (Tajadura-Jiménez and Västfjäll 2008: 66).

The preference for human presence in soundscapes can be associated with our inherent ability to identify animate beings. Based on the results of their neuroscientific inquiry into domain-specific knowledge systems, Caramazza and Shelton argue that mental categories for animate and inanimate objects are products of our evolution and are subserved by dedicated neural mechanisms (1998: 1). These mechanisms have evolved for the categorical representation of animate objects in semantic memory, owing to the survival function of the ability to identify such objects (Vigliocco 2002: B62). These findings suggest that humans display a particular tendency—supported by a dedicated neural system—to semantically differentiate animate beings from inanimate objects.

Controlling the fundamental frequency of vocalizations is central to effective communication among many animals, including humans (Amador and Margoliash 2013: 11136). Gliding pitch variations in intonation can be expressive of not only meaning (Gussenhoven 2002: 47) but also personality and emotion (Scherer and Oshinsky 1977: 332). The gestures consisting of erratic frequency modulations of a sine wave in *Element Yon* were therefore perceived to be suggestive of an organic origin within the abstract sound world of the piece. This was evidenced in descriptors such as "I guess he is trying to tell us something," "communication," "conversation," "crying," and "scream." In response to similar pitch variations in *Birdfish*, many listeners identified birds and cats, on the one hand, and lasers and robots, on the other. Even when the identified source was not organic, it was perceived to engage in conversation. For instance, the participant who imagined a swarm of robotic bugs making progress on a project offered the following general impression: "The dialogues were robotic, but they had emotion. I imagined progress from the emotion in the dialogue."

Contacts between the Two Domains

Given the ceaseless collaboration of perceptual and cognitive faculties in human experience, the physical and the semantic domains of electronic music are intrinsically interwoven. The physical domain constantly informs the semantic domain with new material. In return, the semantic interpretation of gestures contributes to the listener's selective focusing on the material. As the two domains come in contact, the explicit and implicit contexts surrounding the listener become fluid. These kinds of contacts can occur in a number of ways while listening to a piece.

Inside and Outside the Diegesis

The interplay between the semantic and physical attributes of sound implies links between the diegesis and the listening space (e.g., a concert hall). A sound can travel from an alien territory into the concert hall and weave a link between the representational and presentational aspects of the music. A stark example of this phenomenon is observed in Luigi Nono's *La Fabbrica Illuminata*, a 1964 piece for voice and four-channel tape. The piece presents a mixture of live and recorded voices in multichannel audio accompanied by electronic sounds. It's a political piece that depicts the conditions of metalworkers employed at the Italsider foundry in Genoa-Cornigliano and consists of recordings made at this location. The libretto for the piece is drawn from the union meetings and the contract negotiations of the workers (Murphy 2005: 99). The voices on tape transform from quiet speech into loud vocal lines, as they mix with the live singing of a soprano. Quiet sections of the recorded voices create the illusion of a mumbling crowd, which can easily be mistaken for the audience at the concert hall where the piece is being performed. In an interview, Nuria Schönberg-Nono describes that the spatial configuration of the loudspeakers would be adapted to each particular performance space to achieve these kinds of surround effects in the piece. In the same interview, Schönberg-Nono recounts Nono's particular focus on the spatial qualities of auditory phenomena:

> The Basilica of San Marco in Venice was, from his early creative days, a great influence—the idea that music should come from all different directions and that you were in the centre, instead of having all the sound coming to you just from one single source. There are some wonderful films that we have in which he explains these things about how in Venice, when you walk around, you hear so many things coming from different places and he believes that the capacity to hear all these things is in us, but that it has been shut out and it needs to be developed. (Sousa 2008)

In *La Fabbrica Illuminata*, while the performance of the singer embodies a more traditional musical act, it also anchors the piece in the physical domain by serving a presentational function. This amplifies the disorientation when the recordings of the mumbling voices suddenly turn into roaring vocal phrases that are clearly in a space different from that which the audience inhabits. The listener travels back and forth between the concert hall and the foundry. This journey amounts to an eerie experience through the interplay between explicit and implicit worlds. Past voices recorded at the foundry interact with the present voice of the soprano with the listeners situated in between, both physically and metaphorically.

Such physical and semantic qualities of a piece can complement or overpower one another. For instance, the auditory attributes of a sound can manipulate the imagined universe. Conversely, the semantic content of a sound (or lack thereof) can draw attention to its physical characteristics. In *Diegese*, between 0′25″ and 0′50″, during which several sonic layers populate the sound stage, a gesture design inherited from *Birdfish* embodies the former case: the spatial configuration of these organic gestures relative to the concurrent layers causes them to be highly noticeable. While several participants identified these sounds as "bugs" and "insects," one participant submitted "take out these bugs from my ears" as a real-time descriptor. This indicates a distinct embodied engagement with the semantic content of the gestures.

Such interactions will determine where listeners will situate themselves amid the web spun between the physical and the semantic domains. Listeners can witness the unfolding of the musical narrative from inside or outside the diegesis. A participant, who visualized "insects flying in a cave" when listening to *Birdfish*, observed the diegesis from outside. Several participants who listened to *Christmas 2013*, on the other hand, described themselves as the subject of a similar action by expressing such impressions as "flying over a city," "makes you feel as in space," "brings you to the air," "[I imagined] open space, empty or a plain, sea (but still open space behind)." Here, the articulation of space and spatial activity contributes to the listener's sense of immersion, establishing another form of contact between the physical and the semantic domains. In these examples, the listeners allude to an embodied engagement, but one which is contextualized in the diegesis rather than the concert hall. Practically speaking, such impressions can be attributed to the fact that the spatial design of *Christmas 2013* consists of a stable reverberant field and intermittent low-frequency rumbles that were intended to create the illusion of a vast space. But the piece also exploits immediate and intermediate spaces that are constantly articulated with impact sounds traveling *around* the listener. Furthermore, some of the transient gestures display pitch variations reminiscent of Doppler shifts that may have prompted a more convincing sense of a moving sound source. Other participants described their embodied engagement with the piece with phrases such as "inside the brain," "being inside a drum set randomly playing itself," and "music in the air and deep inside the body." All these impressions denote an internalization of the material, which in return causes the participant to situate themselves inside the diegesis. How much of the material is internally (or corporeally) evaluated can therefore dictate the extent to which the listener is inside the implicit world of the piece.

Sense of Time

A narrative is a temporal process. Genette describes that the time needed to consume a narrative is equal to the time needed to traverse it (1980: 34). In literature, this time is borrowed from the pace of reading. In music, the amount of time needed to move through a narrative is set in advance by the composer. As we discussed in Chapter 4, although the progression of time in physical reality displays an empirical consistency, the perceived time ticks in events that can vary in duration. As a result, our understanding of time becomes a function of our "experience of successions" (Fraisse 1963: 1). This implies that the listener's perception of time can also speed up or slow down relative to the unremitting progression of seconds. This temporal relativity constitutes another point of contact between the physical and the semantic domains. The listener inevitably compares the progression of time in their physical reality with that of the spatiotemporal universe implied in a piece. A narrative can therefore be perceived to be accelerating or decelerating as it moves forward.

Participants listening to *Element Yon* referred to a lack of or a slowing down in motion with such general impressions as "something still and stable, not dynamic, not moving" and "slow movement; heaviness." Real-time descriptors such as "heavy," "static," and "waiting" also point to this quality. In *Birdfish*, a participant described in their general impressions how "the sense of time changes over the piece." The same participant used the real-time descriptor "time" after the second movement of the piece commenced. In this section, the ongoing textural density of the piece is pushed into the background. New figure gestures, which exhibit pulsations at slower frequencies, are distributed more sparsely in the foreground. Yet, the event-based pace of the background continues to move forward at the pace of the first movement. This layering was indeed intended to create a contrast between coexisting timescales.

In Chapter 3, I detailed the staccato gesture that serves as a recurring motif in *Christmas 2013*. The individual transients of this gesture display gradually decreasing pitches paired with spatializations that imply a continuous movement for the gesture as a whole. Furthermore, the distance between the individual transients contract and expand in different iterations of the gesture. In combination, these modulations give the sense of an object that is slowing down in the style of an engine coming to a halt or a scene in a movie that goes into slow motion. As a result, various participants referred to their experience of time in this piece with comments such as "trying to stop time by going ultra-slowly," "objects in slow motion," and "the piece made my brain slow down for a moment."

Experienced Listeners

As I mentioned in Chapter 3, the participants of the study came from diverse musical and nonmusical backgrounds. Among the participants were sound engineering and sound art students, who either had been exposed to a repertoire of electronic music or were capable of discussing auditory phenomena in technical terms. A notable tendency in some of the impressions gathered from these participants was the use of *technological listening* (Smalley 1997: 109). This was mostly evident in the *meta descriptors* that addressed the technology or technique behind the music rather than a listener's experience of the music per se. Such impressions ranged from descriptions of synthesis techniques to interpretations of the formal and acoustic properties of the pieces. With these kinds of impressions, the meaning the listener attributes to a sound is influenced by the tools and techniques that underlie its design even when the impression highlights perceptual or semantic properties.

It is worth noting that the feedback from participants who were more seasoned with electronic music and sound art, such as established composers and academics working in these fields, did not rely on technological listening to the same extent. Moreover, the distribution of descriptor categories did not vary notably between experienced and inexperienced listeners in general: while prior exposure seems to have facilitated the characterization of abstract elements, the descriptors that denote representational forms were consistent in terms of both their conceptual disposition and rate of occurrence among participants with contrasting levels of musical experience.

Presence of the Composer in the Work

Earlier on, we talked about how listeners can become conscious of their physical selves while listening to a piece. In a similar vein, sounds can make listeners aware of not just their own embodied presence but also that of the composer. Even in the context of a representationally rich piece, gestures can assume presentational roles that accentuate the presence of a performer. In such pieces, this presence might suggest an *extradiegetic narrator*, which is described as being outside the diegesis but inside the spatiotemporal universe of the *narration* as it relates to the physical world (Bunia 2010: 683).

In the study, reports suggesting the presence of a composer as an extradiegetic narrator manifested in *meta descriptors* that acknowledge a poietic entity behind the work. For instance, in their real-time descriptors, several participants posed questions about the work itself, as in "where is it going?" (*Christmas 2013*), "what's the point?" (*Element Yon*), "is it repeated?" (*Birdfish*), and "repetition, why?" (*Touche pas*). Other descriptors referred to those aspects of the piece that are conceived at the level of narration, such as "final crescendo" (*Birdfish*); "development," "harmonic progression," "motif" (*Touche pas*); "chaotic composition" (*Element Yon*); and "gesture" (*Diegese*). These *meta descriptors* refer to the works as the conscious products of a composition process.

Adorno refers to the aesthetic paradox of "making the impossible possible" when he questions how making can "bring into appearance what is not the result of making" (Adorno 2002: 107). To portray the contrast between showing and telling in a narrative context, Genette proposes a formula where *information + informer = constant*, suggesting an inverse ratio between the quantity of information and the presence of an informer (Genette 1980: 166). In electronic music, the more the composer (i.e., the informer) articulates their presence in the piece, the less of a self-sustaining spatiotemporal universe the piece will evoke. This is an aesthetic preference that can impact the listener's experience substantially. Here, a choice is made between emphasizing the presence of a human performer and maintaining the autonomy of a diegesis or, in other words, between the listener coming into contact with a guiding composer and the listener being left unattended in the world established by the piece. Just as an automated curve controlling the cut-off frequency of a low-pass filter affecting a broadband noise signal can prompt the connotation of a wind or wave, jerky or erratic modulations of the same parameter can easily evoke the image of a performer turning a knob. If it is indeed the intention of an electronic music composer to suggest a spatiotemporal universe entirely separate from that which the listener inhabits, the poietic actions would need to be veiled by the resulting artifact. If the listener can visualize the composer performing a gesture, this will bring about a presentational quality that can overpower the listener's semantic interpretation of the work. This is another powerful narrative device that can be used to shift the emphasis from the semantic domain to the physical domain.

Case Study: *Little Animals*

In this section, I will analyze Natasha Barrett's piece *Little Animals* using some of the theoretical constructs presented so far. This piece, about thirteen minutes in duration, explores many facets of electronic music composition through a mixture of recorded and synthesized sounds that display intricate spectral, temporal, and spatial properties. The piece consists of both abstract and representational elements, the connotative capacity of which transforms over time. Barrett (1999) has also authored a *Computer Music Journal* article that discusses some of the technical and conceptual considerations involved in the composition of this work. In this article, Barrett identifies the most significant aspect of *Little Animals* as the three-part relationship between the listener, the music, and the real-world source inspirations behind the piece. More specifically, the article discusses the interplay between visual and acoustic information that the piece evokes, the transformation of sonic material from musical to

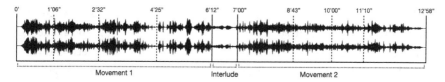

Figure 6.1 Overall structure of *Little Animals* shown on a waveform display of the piece.

extramusical through various sound and space combinations, and the internal and external flow properties that govern the temporal development of the piece.

Macroscale Analysis

Although Barrett's article does not deal with a formal analysis of the piece, it refers to certain macro-level structures that hint at a binary segmentation. For instance, Barrett contrasts the first half of the piece with the second half when she describes the fragmentation of connotatively rich material in the first 6 minutes into pitch textures in the second half (1999: 10). In my macro analysis of this piece, I broke it down to the segments shown in Figure 6.1, which similarly divides the piece into two halves indicated as movements separated by an interlude at the 6'12" mark. While I refrained from labeling the four subsegments within each movement, these roughly adhere to a common narrative arc with introduction, development, and resolution functions.

The opening segment until the 1'06" mark introduces a texture which Barrett characterizes as being evocative of a strong gale (ibid.: 12). This is accompanied by various granular elements that are spread throughout the sound stage with panning and audio decorrelation effects. Halfway through the opening, a resonant line reminiscent of a string instrument introduces the first pitched material in the piece so far. In the following segment between 1'06" and 2'32", the textural density of the piece undulates as new elements are introduced. Among these elements are sounds that indicate an organic origin and those that suggest physical processes, such as objects that are shaken, wound, struck, or dropped on the floor. In this segment, both the time and pitch spaces of the piece expand with new reverberant qualities and tonal threads. A heavy stream of information is then reduced to a high-frequency pattern accompanied by sparse activations of pitched and non-pitched percussive sounds as the tonal thread reaches a cadence in F. Toward the 2'32" mark, low-pass-filtered versions of the existing elements and sub-frequency impulses set a contrasting stage for the following segment where the gestural density of the opening is revisited with objects that go through physical processes like rolling, tumbling, and shattering with a convincing degree of realism. The spatial extent of the piece is further fleshed out with reverberant activations that define a vast space and gestures that move across the panorama. Toward the 4'25" mark, the piece crawls to a halt, at which point the final segment of the first movement begins with the most sporadic distribution of gestures so far. While the objects that indicate physical causalities are still present, some of the pitched elements gain further prominence toward the end of this segment. Without an interruption, an interlude that bridges the first and second movements begins around the 6'12" mark with a sparse arrangement of textures that display periodic patterns on different timescales. Initiating the second movement at the 7'00" mark, some of the gestures from the first movement are reintroduced, but this time masked by low-pass filters and resonators. While repetitive elements in this segment evince an internal flow of time, these are quickly overtaken by the more arrhythmically organized gestures, which, in turn, lend themselves to other droning textures in the segment between 8'43" and 10'00". Subsequently, we are reminded of the objects that adhere to physical causalities, now layered with distant melodic elements. The spatial contrast between these layers is further accentuated as the spatial movement of the objects become disjointed. In the final segment of the piece, we are reminded of a melody that we first encountered toward the end of the first movement around the 5'43" mark. Multiple versions of this melody, created through spectral stretching and pitch shifting, play concurrently as the piece nears to its close. Alongside these melodic lines that travel across the panorama, we hear a distant version of the gale from the opening of the piece.

Gestural Layers

Little Animals employs an extensive sonic vocabulary, comprising a wide range of synthesized and recorded sounds. Whereas some of the synthesized sounds are created with granular techniques, others appear to be the result of physical modeling synthesis. In her article, Barrett does not explicate how these different types of sounds have been assembled into the timeline of the piece. However, the structural heterogeneity of the composition suggests a combination of procedural techniques and micro-montaging.

The representational elements in *Little Animals* conjure up visual imagery that can cause otherwise abstract sound materials to be associated with the real world. In the opening section of the piece, gestures that create the allusion of a strong wind are juxtaposed with a granular noise texture, which does not display obvious representational qualities by itself. However, the coalescence of these two streams amount to a coherent environment, which Barrett characterizes as "the detailed internal life of the forest" (ibid.: 12). Here, the semantic context established by the wind sounds prompts the listener to associate the noise texture with an extramusical concept. Over the course of the first movement, some of the connotative sounds that populate the opening of the piece transform into pitch textures. Amid these textures are what Barrett refers to as "signpost gestures" that help the listener navigate the piece. In the second movement, gestures that display similar acoustic properties as those from the first movement allow extramusical connotations to travel through the piece by way of prompting associations that are intrinsic to the piece (ibid.).

Organic and Environmental Sounds

Barrett's graphic score for *Little Animals* illustrates the gestural and textural unfolding of the piece with inscriptions that indicate pitch and tempo properties and cadences. Furthermore, certain sonic elements in the piece are annotated with object, action, and musical source descriptors such as "organ," "cries," "tearing," and "chimes." In her article, Barrett uses various other source and location descriptors, including "wind," "trees," "roads," and more: "The sound world is rich with a lively animal habitat, and it is likely one will hear the sound of other people walking past, a dog barking, horses' hooves, or a distant tree cutter" (ibid.: 11). As the name of the piece suggests, animal or organic sounds take up a sizeable portion of *Little Animals*' gestural vocabulary; these include glottal sounds, insect sounds, and animal vocalizations, where pitch glides, similar to those used in *Birdfish* and *Element Yon*, convey an organic origin and communicational intent.

In two instances, specifically at the 2′00″ and 5′23″ marks, brief recordings of human voices are used. Although momentary, these instances catch the listener's attention in an unmistakable way. The first recording is introduced in a relatively abstract context, where the pitch contour of the voice is harmonically related to the prevailing tonal thread in the second segment of the first movement. The voice therefore serves a non-diegetic function despite its representational capacity. In the second instance, however, the voice is perceived more as a shouting during a climactic build-up, which creates the illusion of a diegetic actor responding to the events that transpire in the implied universe of the piece.

Physical Causalities

Another prominent gestural layer that persists throughout the piece comprises sounds that indicate physical causalities such as objects colliding, breaking apart, and going through other kinds of mechanical processes. In the macroscale analysis of the piece, I described such sounds as projecting a convincing degree of realism. As a result, this layer often conjures up images of objects and actions situated in the semantic domain. Although these sounds indicate objects of nonorganic origin, the actions they go through allude to beings that set these objects in motion.

In her article, Barrett describes how the fragmentation of the material in the second half of the piece disguises the extramusical associations from the first half, drawing the listener's attention to the intrinsic sonic qualities of the piece (ibid.: 15). This brings about a shift in focus from the semantic domain of the piece to the abstract perceptual attributes of sound. For instance, the gestures that reveal physical causalities in the first movement are reintroduced with prominent resonant filtering in the second half. This causes the said gestures to be perceived as pitched material that contribute to the tonal thread of the piece rather than its semantic context. Barrett describes how enhancing certain frequencies to emphasize pitched content can curtail the acoustic information available for building extramusical associations (ibid.: 14). The opposite is also possible, for instance, when the pitched signpost gesture that marks the end of the first movement comes back at the beginning of the second movement: although the pitch and timbral qualities of this gesture remain the same, its temporal contour follows that of the physical causality gestures from the first half of the piece. As a result, the intrinsic meaning of the gesture is overpowered by a diegetic function.

Pitched and Droning Elements

The tonal thread of the piece consists of sounds that are reminiscent of a musical instrument (e.g., organ, bells), musical phrases that are situated in the implied universe of the piece in the style of diegetic sound, and gestures that are pitched through the use of resonant filters as described above. The spatial treatment of these sounds influences the listener's diegetic contextualization of the associated imagery. Whereas some of these serve abstract musical functions as background layers, others are filtered and reverberated in ways that frame them as parts of the diegesis. For instance, the second movement employs spectral shifting and stretching techniques to morph some of the previously introduced melodic contours into droning and repetitive elements that constitute one of the prominent layers of this movement. These pitched drones are intermixed with non-pitched bubbling and high-frequency textures in dry and reverberated forms. The high-frequency textures, in particular, are less connotative and prompt a degree of perceptual awareness, where the listener grows cognizant of their act of listening.

Besides gestures that are pitched with resonant filters, there are also more abstract pitched elements conceived as signpost gestures that attract the listener's attention. Barrett refers to the use of these gestures as a pitch distraction tactic, where the pitch content of a sound steers the focus away from a concurrent gesture or sound mass with non-pitched elements. These gestures are often detached from the semantic context and serve a more abstract function that foregrounds the internal musical structure of the piece. These distractions elicit an awareness of the separation between the physical domain, wherein the music is *presented* to the listener, and the semantic domain that the extramusical associations establish. A prominent example of this is first introduced at 2'57" with a D-flat impulse that recurs throughout the piece.

Temporal Flow

In her discussion of natural time flow in music, Barrett argues that listeners approach a piece with a set of expectations as to how long an acoustically stimulated object would resonate (ibid.: 14). Therefore, once the listener identifies a sound object, they begin to draw on their knowledge of the environment where such an object would exist. Barrett then distinguishes between internal and external types of flow. Whereas the internal flow of a piece is governed by the intrinsic qualities of the sound material, external flow is rooted in the listener's own sense of timing. When the sound material does not present distinct indicators of internal flow, the listener begins to rely on an external flow that they derive from their memory of the wider musical context. In *Little Animals*, Barrett articulates internal flow with three types of material: widely spaced gestures that demarcate phrases, more closely situated elements that indicate tempo, and periodically accented gestures that reveal metric relationships. While repetitive elements anchor the listener's sense of time in an explicit time flow, the physical causality gestures and the textural elements dispose the listener to rely on their own sense of timing.

The opening segment of the piece is highly animated without a clear rhythmic grid to govern the temporal flow. At the 1'15" mark, the organ sound alludes to a secondary timescale that quickly dissipates. Starting in the third segment of the piece, this organ texture sets the stage for a sparser distribution of gestures, which bring about a slowing down in the internal flow of the piece. Starting at the 5'34" mark, the closing of the first movement introduces a collection of rhythmic patterns, melodic lines, and a pitched signpost gesture, all of which foreground the internal flow of the piece. Although this is disturbed by physical causality gestures that intermittently pull the listener back to an external flow, the second movement more heavily relies on these patterned elements that reinforce the internal flow. In the final two segments of the piece, a layering of rhythmical and arrhythmical elements from the earlier segments allows the listener to choose between the internal and external flows. Subsequently, even the arrhythmical elements are given looping patterns that prioritize the internal flow. The piece concludes with the recap of the ending of the first movement, this time in the absence of diegetic elements that disturb the internal flow of the piece.

Diegetic Disposition of the Listener

In categorizing the sound palette of *Little Animals*, Barrett identifies various types of spaces through which a continuum between musical and extramusical elements can be formed. Whereas a sense of "real space" can be evoked through reverberation and filtering, a sense of "time space" is established based on the listeners' expectations concerning the temporal behavior of physical objects in the real world (ibid.). Barrett describes that the degree of perceptual realism in a piece is a function of the agreement between the realism of a sound and that of the space where this sound is situated. Maximum realism is achieved when a real sound is placed in a recognizable space. Minimum realism, on the other hand, occurs not when an abstract sound is placed in an abstract space but when the sound and space relay conflicting degrees of realism, at which time the plausibility of their combination is diminished. Even when a space suggests a nonreal world, it can still feel familiar to the listener by reminding them of an earlier musical occurrence in the piece. This is reminiscent of Goodman's argument that even abstract works of art can establish world versions by symbolizing their internal features through exemplification.

Little Animals makes substantial use of space with reverberation, filtering, and panning to articulate virtual environments and animated objects, and to establish a relationship between these elements and the listener. At any given time, a multitude of gestures coexist in various spatial configurations ranging from "inside" the listener's head to distances that extend much further beyond the physical domain of the listening experience. Throughout most of the piece, the listener is situated within the diegesis with a first-person view into the implied universe of the piece. The composer establishes this with consistent spatial environments, a sense of physical causality in the animation of gestures, and psychoacoustic effects such as Doppler shifts in gestures that move from one side of the panorama to the other. In combination, these afford the sense of a coherent self-sustaining environment that immerses the listener. The opening of the piece puts the listener at the center of a lively scene where wind-like sounds are foregrounded. Additional layers of transient sounds dispersed across the stereo stage reinforces the listener's contextualization as a part of the diegesis. Starting around the 1′20″ mark, the sound of an organ disturbs this reality: On the one hand, the language and material of this sound gives it the quality of a non-diegetic score separate from the representational environment introduced so far. On the other hand, the reverberant properties of the sound maintain a link to the diegesis, suggesting that it may be a distant element of the environment the listener is immersed in. At the 1′40″ mark, we hear Doppler shifts in sounds that move from one ear to the other, creating a strong illusion of vehicles passing by. In the following segment starting at 2′30″, the sonic environment begins to exhibit the reverberant properties of a small- to medium-sized room, establishing a more intimate context where the foreground elements feel closer to the listener. While the reverberant space of the organ sound remains distant, the semantic ambiguity of this layer is further amplified as its temporal contour begins to react to changes in the foreground gestures. Here, a contact is established between the diegetic actors and the musical elements that are abstracted from the diegesis. Some of the tearing sounds in the foreground become dry to the extent that they evoke an in-the-head sensation that draws the listener's attention to their own hearing while the tonal thread is pushed further away as a separate diegetic layer that the listener observes from afar. The D-flat signpost gesture at the 2′57″ mark is the most overtly non-diegetic element introduced thus far in the piece, but this sound is not detached from the unfolding of the events in the foreground layers either. A sub-bass activation at the 3′40″ mark creates a moment of embodied engagement followed by a flurry of physical causality gestures. The silence around 4′20″ allows the listener to pay attention to their physical space; the sonic material is reduced to a dry gesture that, despite its spatial animation, feels like a part of the listener's immediate surroundings. This is quickly contrasted by distant sounds that are suggestive of the diegesis. Another bass relief at the 4'40" mark, although spectrally similar to the previous one, now articulates a reverberant space, therefore shifting the focus further toward the semantic domain. A build-up between 5'20" and 5'26" resolves into the aforementioned recording of a shouting voice. Another build-up in half a minute leads to the loudest incarnation of the D-flat signpost gesture that acts almost like a non-diegetic counterpart of the human voice as the first movement nears to its conclusion. The following interlude serves a more presentational function with a sparse distribution of abstract musical elements and clear rhythmic patterns.

In the second movement of the piece the gestural density picks back up, but with an amplification of pitch content through resonant filtering. The segment changes in this movement are marked by shifts in the tonal thread of the piece. As pitch becomes a growingly definitive aspect of the piece, interesting interactions between the physical and the semantic domains occur. For instance, the reverberant qualities of a melodic line introduced at the 10'00" mark gives the impression of a small room. When paired with the low-pass filtering applied to this line, the virtual space affords the sense of a melody playing behind a wall. Here, a melodic line that would normally be devoid of an extramusical connotation is given a diegetic quality, creating the impression that it is heard "next door" to the diegetic space that the listener is situated in. This contrast is further articulated with dry and unfiltered gestures that reinforce the sense of an immediate space surrounding the listener. Alongside these two layers, other gestures near and far define additional reverberant spaces: the reverberant response to an up-close high-frequency impulse reveals that the space surrounding the listener might be more expansive than previously assumed; low-frequency impulses at a distance support this implied extent of the environment. When paired with the dry sounds and the small-room reverberation, these elements begin to suggest an implausibility of the diegesis. This implausibility is further accentuated starting at the 11'10" mark when the piece introduces the first instances of a gesture whose reverberant tail gets cut off abruptly, breaking the causal consistency of the space and exaggerating the notion that the implied universe of the piece is not "real." This is followed by other foreground gestures being similarly cut off before they reach their conclusion in an acoustically convincing way. Here, the presence of the composer (i.e., someone manipulating the implied universe of the piece) gains a narrative function in the style of a *reveal* plot device. With the self-sustaining nature of the diegesis disrupted, we discover a poietic entity behind the world of *Little Animals*.

Overall, *Little Animals* is a powerful demonstration of how a piece of electronic music can traverse the continuum between abstractness and representationality. Over the course of thirteen minutes, the piece introduces numerous layers that stretch across the physical and the semantic domains of the listening experience. While the concurrency of these layers often affords the listener an esthesic license to prioritize one domain over the other, the composer introduces elements that play a strategic role in swaying the listener's attention to a specific domain. The interplay between the two and the intricate ways in which the diegetic properties of certain gestures transform over time play a structural role in the narrative unfolding of the piece; these not only influence the visual imagery associated with the sonic material but also articulate the formal structure of the piece by way of demarcating its segments.

Coda

In her book *Listening through the Noise*, Joanna Demers points out how artists have little time to contemplate why electronic music is ontologically different from other kinds of music (2010: 5). To some degree, the study presented here allowed me to delegate this contemplation to listeners. Through a combination of post-hoc and ad-hoc impressions of a piece of electronic music, the participants conveyed how they engage with this music at both conceptual and perceptual levels and on micro and macro scales of form. These impressions were highly diverse and, in many cases, reflective of the participants' life experiences. Some of them associated what they heard with childhood memories, whereas others offered distinctly individualistic accounts of what they felt, thought, or imagined while they listened to the pieces that were presented to them. Some have expressed themselves through stories, while others made illustrations. The thematic analysis of all this feedback was an unwieldy task, but, with the support of my peers, I parsed through thousands of descriptors several times over to devise and refine the general categories presented in Chapter 3. These results bared similarities with not only existing taxonomies of electronic music but also categorizations that are derived from experimental studies on auditory cognition and environmental sounds. Throughout the book, I attempted to build a web across these theoretical and experimental perspectives and the study presented here. I anchored this conversation with technical details of the sound designs and compositions that prompted the responses I gathered from the listeners. I synthesized philosophies of art and aesthetics with models of perception to convey these results in the form of an interdisciplinary discourse. I hope that these efforts can stimulate new conversations among a range of artistic and academic communities.

As electronic music practices throughout history have demonstrated, research is often an intrinsic aspect of creativity in this domain. Many composers explore the fringes of technology and perception and, in doing so, allow technology and perception to govern their artistic practices. This book offers an account of my research through practice, leveraging the interplay between the electronic medium and auditory cognition in an artistic context. Despite the rigor applied to the design and analysis of the study presented here, I acknowledge that quantifying an aesthetic experience can hardly be reduced to an exact science. I cannot, for instance, argue that the results offer a definitive representation of listeners' perception of electronic music. Moreover, these results are by no means intended to suggest a "better" way of composing music. They do, however, highlight some of the unique ways in which we conceptualize this music, prioritize certain aspects of it, and relate it to our own selves. The results also reflect a dialogue between poietic techniques and intents, on the one side, and the complexity of listening, on the other. That being said, as with most experimental studies, these results would benefit from a wider pool of participants. I relayed the study procedure and its results with as much transparency and detail as possible to enable other artists and researchers to replicate or modify this study and expand its scope with not only more participants but also other works and styles of electronic music. Furthermore, recent advances in cognitive science, particularly in artificial intelligence and the modeling of cognitive phenomena, can offer new ways to interpret the results obtained from studies like this, and allow the design of new compositional and performative systems based on such results.

While the book primarily focuses on electronic music, the application domain of the findings and theories presented here can be much wider. For instance, beyond the idiomatic uses of the electronic gesture as an analytical device in music, the concepts of causality, narrativity, and intentionality can facilitate new discussions into the phenomenology of art, with a particular emphasis on the parallels between electronic music and digital media arts, where similar considerations of abstractness, representationality, and technological mediation come into play. From a more practical standpoint, among the areas that could make use of the findings presented here are sound design, game audio, and film music. Creative efforts in these domains are often aimed at eliciting specific affective and cognitive reactions from audiences. The pairings between the technical discussion of gestures and listener responses presented in Chapters 4, 5, and 6 could serve as sound design strategies for practitioners working in these areas. Furthermore, critical studies in these domains could benefit from the theoretical constructs presented here. For instance, the concept of diegetic affordances discussed in Chapter 5 and the domains of listening experience explored in Chapter 6 are directly correlated with theories on diegetic and non-diegetic sound as they pertain to sound for visual media.

A related field that has seen significant growth in recent years is virtual reality (VR). With the introduction of more accessible hardware and software tools, VR has quickly turned into a new frontier for creative expression. Many artists have already begun to exploit the unique affordances of VR for musical creativity, and such practices are bound to proliferate as both VR technology itself and our grasp of VR as a creative medium advance. This is an area that I have been engaged in both artistically and academically for over a decade. The immersive contexts and the first-person agency that VR offers to its users in arbitrarily defined virtual systems have many implications for musical practices as well. In that regard, the perspectives on worldmaking and diegesis presented in Chapter 5 can inform studies in this domain both conceptually and practically. The question of where the listener is situated in relation to the implied universe of a piece takes on an explicit meaning in VR, setting off broader inquiries into the intersection between music and virtuality and the allocation of creative license to the listener.

Early on in the book, I characterized electronic music as a powerful form of artistic expression that can evoke a diverse range of experiences. It can indeed diverge from what is conventionally regarded as music, and it may require a bit of "getting used to." But, as scientific evidence suggests, so does all music. Various studies cited throughout the book have shown that our appreciation of music is socioculturally conditioned. We learn to appreciate music, and the extent to which we are exposed to a certain kind of music has an influence on what we like about this art form. In Chapter 1, I described instances where electronic music styles that sprouted from avant-garde practices evolved or assimilated into mainstream genres. As a result of such evolutions, our vocabulary of musical sounds has greatly expanded over the course of the past century. Yet electronic music still has much wider audiences to reach. During his seminal lecture at the Oxford Union in 1972, Stockhausen was asked whether the music we are brought up on is a barrier to the appreciation of this music and what this implies for the education of the public. Stockhausen responded, "Well, I have six children and they pick it up as the most natural thing in the world." This anecdote nicely illustrates how getting exposed to different kinds of music, especially in childhood, can play a significant role in the development of our musical taste. I believe the further we understand the electronic music experience and acknowledge the multisensory context that it is anchored in, the more easily we can formulate strategies to make this music reach broader audiences at an earlier age.

In Chapter 1, I empathized with the early-twentieth-century audiences by questioning whether what music was evolving into back then was still music. Such a question continues to be raised in response to many flavors of electronic music to this day: *Is this music?* Weeding out the instances that were intended to spark an ontological debate about what music is, I recall encountering this question in various forms throughout my artistic career. I greatly appreciate this question since it is infinitely more promising than an unhesitating dismissal. It implies that an uninitiated listener can be open to the idea of accepting what they have heard as music. Oftentimes, this question is compounded with an enthusiastic description of what the listener thought was happening with the piece they just heard. The kinds of curiosity and fascination that listeners express in response to this music is not unlike those that urged the artists in the early twentieth century to experiment with electronic sound in the first place. Today, audio recording, synthesis, and processing tools are not only more capable but also more accessible than they have ever been. As a result, we are witnessing a new era of sonic exploration, where people from all segments of the population engage in electronic music–making and share their work with a wide range of communities through social media and streaming services. As our fascination with sound takes on a growingly participatory form, our notions of music become more amorphous and inclusive. These are qualities that have defined electronic music since its early days and continue to shape our creative explorations in the electronic medium today.

The listening experiences that electronic music offers permeate our daily environments, which are replete with auditory events that show complex spectral, dynamic, and spatial properties. The mental mechanisms with which we parse these environments come into play when we listen to electronic music as well. In my introduction to this book, I argued that electronic music engages with listening abilities that we take for granted in our everyday lives and show us how intricate they can be. The study presented here has shown that electronic music can make us feel emotions, become aware and unaware of our physical selves, and use our imaginations in unique and personal ways. I hope that this book expands our comprehension of the experiential depth that this music affords. There is much more to be gained from electronic music, and its ability to stimulate our cognitive faculties in such diverse ways is one of its most remarkable offerings.

Acknowledgments

This book is based on research that I began to carry out at the Leiden University, Academy of Creative and Performing Arts, in affiliation with the Institute of Sonology at the Royal Conservatory of The Hague and the Faculty of Industrial Design Engineering at the Delft University of Technology. First, I would like to thank my advisors Vincent Meelberg, Elif Özcan, Richard Barrett, and Frans de Ruiter for their invaluable support and guidance. My work was reviewed by a group of field experts, including Clarence Barlow, Marcel Cobussen, Nic Collins, Simon Emmerson, Bob Gilmore, and Larry Polansky. I am thankful for their input, which has helped me improve my research and, ultimately, this book. I owe a debt of gratitude to my former teacher Curtis Roads, who has not only provided valuable feedback during this research but also allowed me to use one of his unpublished works in the study presented here. I would also like to thank my colleague Mary Francis for sharing her vast wisdom of books with me. I appreciate all the support I received from my students, peers, and colleagues at the Istanbul Technical University, Leiden University, University of Illinois at Chicago, and University of Michigan while I prepared this book. I am especially thankful to my colleagues at the University of Michigan's Department of Performing Arts Technology and the broader School of Music, Theatre & Dance for the artistically and intellectually vibrant environment they fostered as I worked on this book these past few years. I am also grateful to all eighty participants of the study for generously sharing their impressions with me. Last but not least, I would like to thank my spouse, Zeynep Özcan; my parents, Meral and Akın Çamcı; and my brother, Alp Çamcı; this book would not have been possible if it weren't for their unwavering patience and support over the years.

Bibliography

Adorno, T. W. 2002. *Aesthetic Theory*. Gretel Adorno and Rolf Tiedemann (eds.). Translated by Robert Hullot-Kentor. Continuum, London.

Adorno, T. W., and Gillespie, S. 1993. "Music, Language, and Composition." *Musical Quarterly* 77(3): 401–14.

Alexenberg, M. 2004. "Semiotic Redefinition of Art in the Digital Age." In Debbie Smith-Shank (ed.), *Semiotics and Visual Culture: Sights, Signs, and Significance*: 124–31. National Art Education Association, Reston, VA.

Amacher, M. 2008. "Psychoacoustic Phenomena in Musical Composition: Some Features of a 'Perceptual Geography.'" In John Zorn (ed.), *Arcana III: Musicians on Music*: 9–24. Hips Road, New York.

Amador, A., and Margoliash, D. 2013. "A Mechanism for Frequency Modulation in Songbirds Shared with Humans." *Journal of Neuroscience* 33(27): 11136–44.

Ashline, W. L. 1995. "The Problem of Impossible Fictions." *Style* 29(2): 215–34.

Bahn, C., Hahn, T., and Trueman, D. 2001. "Physicality and Feedback: A Focus on the Body in the Performance of Electronic Music." *Proceedings of the International Computer Music Conference 2001*, Havana, Cuba.

Bal, M. 1997. *Narratology: Introduction to the Theory of Narrative*, 2nd ed. University of Toronto Press, Toronto, Canada.

Ballantine, C. 1979. "Charles Ives and the Meaning of Quotation in Music." *Musical Quarterly* 65(2): 167–84.

Ballas, J. A. 1993. "Common Factors in the Identification of an Assortment of Brief Everyday Sounds." *Journal of Experimental Psychology: Human Perception and Performance* 19(2): 250–67.

Ballas, J. A., and Howard Jr., J. H. 1987. "Interpreting the Language of Environmental Sounds." In *Environment and Behavior* 19(1): 91–114.

Ballas, J. A., and Mullins, T. 1991. "Effects of Context on the Identification of Everyday Sounds." *Human Performance* 4(3): 199–219.

Bar, M. 2004. "Visual Objects in Context." *Nature Reviews, Neuroscience* 5(8): 617–29.

Barrett, N. 1999. "Little Animals: Compositional Structuring Processes." *Computer Music Journal* 23(2): 11–18.

Barrett, R. 2012. Personal communication on May 29.

Barrett, R. 2013. Personal communication on September 13.

Barsalou, L. W. 1999. "Perceptual Symbol Systems." *Behavioral and Brain Sciences* 22(4): 557–609.

Bartlett, J. C. 1977. "Remembering Environmental Sounds: The Role of Verbalization at Input." *Memory & Cognition* 5(4): 404–14.

Basanta, A. 2013. "Tracing Conceptual Structures in Listener Response Studies." *eContact!* 15(2). Accessed July 2, 2014, at http://cec.sonus.ca/econtact/15_2/basanta_listenerresponse.html.

Bayle, F. 1993. *Musique Acousmatique—Propositions ... Positions*. Buchet/Chastel, Paris.

Beaudoin, R. 2007. "Counterpart and Quotation in Ussachevsky's *Wireless Fantasy*." *Organised Sound* 12(2): 143–51.

Bertelsen, L., and Murphie, A. 2010. "An Ethics of Everyday Infinities and Powers: Félix Guattari on Affect and the Refrain." In Melissa Gregg and Gregory J. Seigworth (eds.), *The Affect Theory Reader*: 138–60. Duke University Press, Durham, NC.

Beta, A. 2016. "Blood and Echoes: The Story of Come Out, Steve Reich's Civil Rights Era Masterpiece." *Pitchfork*. Accessed October 9, 2019, at https://pitchfork.com/features/article/9886-blood-and-echoes-the-story-of-come-out-steve-reichs-civil-rights-era-masterpiece/.

Bhatara, A., Tirovolas, A. K., Duan, L. M., Levy, B., and Levitin, D. J. 2011. "Perception of Emotional Expression in Musical Performance." *Journal of Experimental Psychology: Human Perception and Performance* 37(3): 921–34.

Blacking, J. 1973. *How Musical Is Man?* University of Washington Press, Seattle, WA.

Boulez, P. 1986. "Technology and the Composer." In Simon Emmerson (ed.), *The Language of Electroacoustic Music*: 5–14. Macmillan Press, Houndmills, UK.

Bradley, M. M., and Lang, P. J. 2000. "Affective Reactions to Acoustic Stimuli." *Psychophysiology* 37(2): 204–15.

Brazil, E., Fernström, M., and Bowers, J. 2009. "Exploring Concurrent Auditory Icon Recognition." *Proceedings of the 15th International Conference on Auditory Display*, May 18–22, Copenhagen, Denmark.

Bridger, M. 1989. "An Approach to the Analysis of Electro-acoustic Music Derived from Empirical Investigation and Critical Methodologies of Other Disciplines." *Contemporary Music Review* 3(1): 145–60.

Bridger, M. 1993. "Narrativisation in Electroacoustic and Computer Music—Reflections on Empirical Research into Listeners' Response." *Proceedings of the International Computer Music Conference 1993*, Tokyo, Japan: 296–99.

Brown, S., Merker, B., and Wallin, N. L. 2001. "An Introduction to Evolutionary Musicology." In Nils L. Wallin, Björn Merker, and Steven Brown (eds.), *The Origin of Music*. MIT Press, Boston.

Bunia, R. 2010. "Diegesis and Representation: Beyond the Fictional World, on the Margins of Story and Narrative." *Poetics Today* 31(4): 679–720.

Burks, A. W. 1949. "Icon, Index, and Symbol." *Philosophy and Phenomenological Research* 9(4): 673–89.

Cadoz, C., and Wanderley, M. M. 2000. "Gesture—Music." In Marcelo M. Wanderley and Marc Battier (eds.), *Trends in Gestural Control of Music*: 71–94. Ircam-Centre Pompidou, Paris.

Cage, J. 2004. "The Future of Music: Credo." In Christopher Cox and Daniel Warner (eds.), *Audio Culture: Readings in Modern Music*: 25–8. Continuum, New York.

Çamcı, A. 2012a. "Christmas 2013." Program notes. The Global Composition: World Soundscape Conference, July 25–28, Dieburg, Germany.

Çamcı, A. 2012b. "A Cognitive Approach to Electronic Music: Theoretical and Experiment-Based Perspectives." *Proceedings of the 38th International Computer Music Conference (ICMC), September 2012*, Ljubljana, Slovenia: 1–4.

Çamcı, A. 2013. "Diegesis as a Semantic Paradigm for Electronic Music." *eContact!* 15(2). Accessed June 15, 2013, at https://econtact.ca/15_2/camci_diegesis.html.

Çamcı, A. 2016. "Imagining through Sound: An Experimental Analysis of Narrativity in Electronic Music." *Organised Sound* 21(3): 179–91.

Çamcı, A., Çakmak, C., and Forbes, A. G. 2019. "Applying Game Mechanics to Networked Music HCI Systems." In Simon Holland, Tom Mudd, Katie Wilkie-McKenna, Andrew McPherson, and Marcelo M. Wanderley (eds.), *New Directions in Music and Human-Computer Interaction*: 223–41. Springer, Cham, Germany.

Çamcı, A., and Erkan K. 2013. "Interferences between Acoustic Communication Threads in Enclosed Social Environments of Istanbul." *Soundscape: Journal of Acoustic Ecology* 12(1): 20–4.

Çamcı, A., and Meelberg, V. 2016. "Diegetic Affordances and Affect in Electronic Music." *Proceedings of the 42nd International Computer Music Conference* (ICMC), September 2016, Utrecht, the Netherlands: 104–9.

Çamcı, A., and Özcan, E. 2018. "Comparing the Cognition of Abstract and Representational Structures in Electronic Music." *Proceedings of the 15th International Conference on Music Perception and Cognition* (ICMPC), July, Graz, Austria.

Çamcı, A., Vilaplana, M., and Wang, R. 2020. "Exploring the Affordances of VR for Musical Interaction Design with VIMEs." *Proceedings of the International Conference of New Interfaces for Musical Expression* (NIME), July 2020, Birmingham.

Caramazza, A., and Shelton, J. R. 1998. "Domain-Specific Knowledge Systems in the Brain: The Animate-Inanimate Distinction." *Journal of Cognitive Neuroscience* 10(1): 1–34.

Chadabe, J. 1997. *Electric Sound: The Past and Promise of Electronic Music*. Prentice Hall, Upper Saddle River, NJ.

Chapin, H., Jantzen, K., Kelso, J. S., Steinberg, F., and Large, E. 2010. "Dynamic Emotional and Neural Responses to Music Depend on Performance Expression and Listener Experience." *PLoS ONE* 5(12): 1–14.

Chion, M. 1994. *Audio-Vision: Sound on Screen*. Edited and translated by Claudia Gorbman. Columbia University Press, New York.

Chowning, J. M. 1971. "The Simulation of Moving Sound Sources." *Journal of the Audio Engineering Society* 19(1): 2–6.

Chowning, J. M. 2008. "Fifty Years of Computer Music: Ideas of the Past Speak to the Future." In Richard Kronland-Martinet, Sølvi Ystad, Kristoffer Jensen (eds.), *Computer Music Modeling and Retrieval: Sense of Sounds*, LNCS 4969: 1–10. Springer, Berlin.

Ciciliani, M. 2014. "Towards an Aesthetic of Electronic-Music Performance Practice." *Proceedings of the Joint International Computer Music Conference and Sound and Music Conference 2014*, September 14–20, Athens, Greece: 262–8.

Coker, W. 1972. *Music and Meaning: A Theoretical Introduction to Musical Aesthetics*. Free Press, New York.

Collins, N. 2010. *Introduction to Computer Music*. Wiley, Hoboken, NJ.

Coward, S. W., and Stevens, C. J. 2004. "Extracting Meaning from Sound: Nomic Mappings, Everyday Listening, and Perceiving Object Size from Frequency." *Psychological Record* 54(3): 349–64.

Cross, I. 2010. "The Evolutionary Basis of Meaning in Music: Some Neurological and Neuroscientific Implications." In Frank Clifford Rose (ed.), *The Neurology of Music*: 1–15. Imperial College Press, London.

Cummings, A., Čeponienė, R., Koyama, A., Saygin, A. P., Townsend, J., and Dick, F. 2006. "Auditory Semantic Networks for Words and Natural Sounds." *Brain Research* 1115(1): 92–107.

Curtis, M. E., and Bharucha J. J. 2010. "The Minor Third Communicates Sadness in Speech, Mirroring Its Use in Music." *Emotion* 10(3): 335–48.

Dack, J. 1999. "Karlheinz Stockhausen's Kontakte and Narrativity." *eContact!* 2(2). Accessed May 10, 2010, at http://cec.sonus.ca/econtact/SAN/Dack.htm.

Delalande, F. 1998. "Music Analysis and Reception Behaviours: *Sommeil* by Pierre Henry." *Journal of New Music Research* 27(1–2): 13–66.

Dalgleish, M. 2016. "The Modular Synthesizer Divided: The Keyboard and Its Discontents." *eContact!* 17(4). Accessed June 12, 2020, at https://econtact.ca/17_4/dalgleish_keyboard.html.

Deleuze, G. 1978. *Continuous Variation, Lecture 12, 24 January 1978*. Translated by Timothy S. Murphy. Seminar at the University of Paris, Vincennes-St. Denis, 1977–8.

Deleuze, G. 2003. *Francis Bacon: The Logic of Sensation*. Translated by Daniel W. Smith. Continuum, London.

Deleuze, G., and Guattari, F. 1987. *A Thousand Plateaus: Capitalism and Schizophrenia*. Translated by Brian Massumi. University of Minnesota Press, Minneapolis.

Deleuze, G., and Guattari, F. 2000. "Percept, Affect, and Concept." In Clive Cazeaux (ed.), *The Continental Aesthetics Reader*: 465–88. Routledge, New York.

Demers, J. 2010. *Listening through the Noise: The Aesthetics of Experimental Electronic Music*. Oxford University Press, New York.

Depew, D. J. 2003. "Baldwin and His Many Effects." In Bruce H. Weber and David J. Depew (eds.), *Evolution and Leaning: The Baldwin Effect Reconsidered*: 3–31. MIT Press, Cambridge, MA.

d'Escriván, J. 2006. "To Sing the Body Electric: Instruments and Effort in the Performance of Electronic Music." *Contemporary Music Review* 25(1–2): 183–91.

Deutsch, D. 1980. "The Processing of Structured and Unstructured Tonal Sequences." *Perception and Psychophysics* 28(5): 381–9.

Doornbusch, P. 2011. "Appendix: A Chronology of Computer Music and Related Events." In Roger T. Dean (ed.), *The Oxford Handbook of Computer Music*: 557–607. Oxford University Press, New York.

Dubois, D. 2000. "Categories as Acts of Meaning: The Case of Categories in Olfaction and Audition." *Cognitive Science Quarterly* 1(1): 35–68.

Dubois, D., Guastavino, C., and Raimbault, M. 2006. "A Cognitive Approach to Urban Soundscapes: Using Verbal Data to Access Everyday Life Auditory Categories." *Acta Acustica United with Acustica* 92(6): 865–74.

Eitan, Z., and Granot, R. Y. 2006. "How Music Moves: Musical Parameters and Listeners' Images of Motion." *Music Perception* 23(3): 221–47.

Ekkekakis, P. 2012. "Affect, Mood, and Emotion." In Gershon Tenenbaum, Robert C. Eklund, and Akihito Kamata (eds.), *Measurement in Sport and Exercise Psychology*: 321–32. Human Kinetics, Champaign, IL.

Elgin, C. Z. 2001. "The Legacy of Nelson Goodman." *Philosophy and Phenomenological Research* 62(3): 679–90.

Emmerson, S. 1986. "The Relation of Language to Materials." In Simon Emmerson (ed.), *The Language of Electroacoustic Music*: 17–40. Macmillan Press, Houndmills, UK.

Feaster, P. 2019. "Enigmatic Proofs: The Archiving of Édouard-Léon Scott de Martinville's Phonautograms." *Technology and Culture* 60(2): S14–S38.

Field, A. 2000. "Simulation and Reality: The New Sonic Objects." In Simon Emmerson (ed.), *Music, Electronic Media and Culture*: 36–55. Ashgate, Burlington, VT.

Fitch, W. T. 2006. "The Biology and Evolution of Music: A Comparative Perspective." *Cognition* 100(1): 173–215.

Fraisse, P. 1963. *The Psychology of Time*. Harper & Row, New York.

Frisk, H., and Östersjö, S. 2006. "Negotiating the Musical Work. An Empirical Study on the Inter-relation between Composition, Interpretation and Performance." *Proceedings of Electroacoustic Music Studies Network (EMS) Conference 2006*, Beijing, China.

Gaver, W. W. 1993a. "What in the World Do We Hear? An Ecological Approach to Auditory Source Perception." *Ecological Psychology* 5(1): 1–29.

Gaver, W. W. 1993b. "How Do We Hear in the World? Explorations in Ecological Acoustics." *Ecological Psychology* 5(4): 285–313.

Genette, G. 1969. "D'un Récit Baroque." In *Figures II*: 194–222. Éditions du Seuil, Paris.
Genette, G. 1980. *Narrative Discourse*. Translated by Jane E. Lewin. Cornell University Press, New York.
Gerrig, R. J., and Egidi, G. 2003. "Cognitive Psychological Foundations of Narrative Experiences." In David Herman (ed.), *Narrative Theory and the Cognitive Sciences*: 33–55. Center for the Study of Language and Information, Stanford, CA.
Gibson, J. J. 1963. "The Useful Dimensions of Sensitivity." *American Psychologist* 18(1): 1–15.
Gibson, J. J. 1966. *The Senses Considered as Perceptual Systems*. Greenwood Press, Westport, CT.
Gibson, J. J. [1979] 1986. *The Ecological Approach to Visual Perception*. Psychology Press, New York.
Gjerdingen, R. O. 2013. "Psychologists and Musicians: Then and Now." In Diana Deutsch (ed.), *The Psychology of Music*: 683–707. Academic Press, Cambridge, MA.
Godøy, R. I. 2006. "Gestural-Sonorous Objects: Embodied Extensions of Schaeffer's Conceptual Apparatus." *Organised Sound* 11(2): 149–57.
Goldstein, E. B. 1981. "The Ecology of J. J. Gibson's Perception." *Leonardo* 14(3): 191–5.
Goodman, N. 1968. *Languages of Art: An Approach to a Theory of Symbols*. Bobbs-Merrill, Indianapolis.
Goodman, N. 1978. *Ways of Worldmaking*. Hackett, Indianapolis.
Gorbman, C. 1980. "Narrative Film Music." *Yale French Studies* 60: 183–203.
Grainger, P. A. 1996. "Free Music." *Leonardo Music Journal* 6: 109.
Gritten, A., and King, E. 2006. "Introduction." In Anthony Gritten and Elaine King (eds.), *Music and Gesture*: xix–xxv. Ashgate, Burlington, VT.
Guastavino, C. 2007. "Categorization of Environmental Sounds." *Canadian Journal of Experimental Psychology* 61(1): 54–63.
Gussenhoven, C. 2002. "Intonation and Interpretation: Phonetics and Phonology." *Proceedings of the First International Conference on Speech Prosody*, April 11–13, Aix-en-Provence, France: 47–57.
Gygi, B., Kidd, G. R., and Watson, C. S. 2004. "Spectral-Temporal Factors in the Identification of Environmental Sounds." *Journal of the Acoustical Society of America* 115(3) March: 1252–65.
Gygi, B., Kidd, G. R., and Watson, C. S. 2007. "Similarity and Categorization of Environmental Sounds." *Perception & Psychophysics* 69(6): 839–55.
Hatten, R. S. 2003. *Musical Gesture: Theory and Interpretation*. Course description, Indiana University. Accessed April 12, 2014, at https://web.archive.org/web/20140607185945/http://www.indiana.edu/~deanfac/blfal03/mus/mus_t561_9824.html.
Hatten, R. S. 2006. "A Theory of Music Gesture and Its Application to Beethoven and Schubert." In Anthony Gritten and Elaine King (eds.), *Music and Gesture*: 1–4. Ashgate, Burlington, VT.

Hayward, P. 1997. "Danger! Retro-Affectivity! The Cultural Career of the Theremin." *Convergence* 3(4): 28–53.

Hayward, S. 2006. *Cinema Studies: The Key Concepts*, 3rd ed. Routledge, New York.

Heathcote, A. 2003. "Liberating Sounds: Philosophical Perspectives on the Music and Writings of Helmut Lachenmann." Doctoral dissertation, Durham University, UK.

Herman, D. 2009. *Basic Elements of Narrative*. Wiley-Blackwell, West Sussex, UK.

Hill, A. 2013. "Understanding Interpretation, Informing Composition: Audience Involvement in Aesthetic Result." *Organised Sound* 18(1): 43–59.

Holm-Hudson, K. 1997. "Quotation and Context: Sampling and John Oswald's Plunderphonics." *Leonardo Music Journal* 7: 17–25.

Holmes, T. 2008. *Electronic and Experimental Music, Technology, Music and Culture*, 3rd ed. Routledge, New York.

Howard, D. 2008. "From Ghetto Laboratory to the Technosphere: The Influence of Jamaican Studio Techniques on Popular Music." *Proceedings of the 4th Art of Record Production Conference*, Lowell, MA.

Howe, H. S. 1972. "Compositional Limitations of Electronic Music Synthesizers." *Perspectives of New Music* 10(2): 120–9.

Huron, D. 2006. *Sweet Anticipation, Music and the Psychology of Expectation*. MIT Press, Cambridge, MA.

Hutton, J. 2003. "Daphne Oram: Innovator, Writer and Composer." *Organised Sound* 8(1), 49–56.

Iazzetta, F. 2000. "Meaning in Musical Gesture." In Marcelo M. Wanderley and Marc Battier (eds.), *Trends in Gestural Control of Music*: 259–68. Ircam-Centre Pompidou, Paris.

Ihde, D. 2007. *Listening and Voice: Phenomenologies of Sound*. SUNY Press, Albany, NY.

Jackendoff, R. 2009. "Parallels and Nonparallels between Language and Music." *Music Perception* 26(3): 195–204.

Jackendoff, R., and Lerdahl, F. 2006. "The Capacity for Music: What Is It, and What's Special about It?" *Cognition*: 100(1): 33–72.

Jekosch, U. 2005. "Assigning Meaning to Sounds—Semiotics in the Context of Product-Sound Design." In Jens Blauert (ed.), *Communication Acoustics*: 193–221. Springer, Berlin.

Jenkins, J. J. 1985. "Acoustic Information for Objects, Places, and Events." In William H. Warren Jr. and Robert E. Shaw (eds.), *Persistence and Change: Proceedings of the First International Conference on Event Perception*: 115–39. Lawrence Erlbaum, Hillsdale, NJ.

Jensenius, A. R. 2007. "ACTION—SOUND: Developing Methods and Tools to Study Music-Related Body Movement." Doctoral dissertation, Department of Musicology, University of Oslo, Norway.

Juslin, P. N. 2001. "Communicating Emotion in Music Performance: A Review and a Theoretical Framework." In Patrik N. Juslin and John A. Sloboda (eds.), *Music and Emotion: Theory and Research*: 309–37. Oxford University Press, New York.

Juslin, P. N., and Västfjäll, D. 2008. "Emotional Responses to Music: The Need to Consider Underlying Mechanisms." *Behavioral and Brain Sciences* 31: 559–621.

Kane, B. 2007. "L'Objet Sonore Maintenant: Pierre Schaeffer, Sound Objects and the Phenomenological Reduction." *Organised Sound* 12(1): 15–24.

Kane, B. 2014. *Sound Unseen: Acousmatic Sound in Theory and Practice*. Oxford University Press, New York.

Karbusický, V. 1969. "Electronic Music and the Listener/La Musique Electronique et les Auditeurs/Elektronische Musik und Hörer." *World of Music* 11(1), 32–44.

Katz, M. 2001. "Hindemith, Toch, and Grammophonmusik." *Journal of Musicological Research* 20(2): 161–80.

Kendall, G. S. 2010. "Meaning in Electroacoustic Music and the Everyday Mind." *Organised Sound* 15(1): 63–74.

Koelsch, S., Kasper, E., Sammler, D., Schulze, K., Gunter, T., and Friederici, A. D. 2004. "Music, Language and Meaning: Brain Signatures of Semantic Processing." *Nature Neuroscience* 7(3): 302–7.

Kramer, J. D. 1978. "Moment Form in the Twentieth Century Music." *Musical Quarterly* 64(2): 177–94.

Krause, B. 2012. *The Great Animal Orchestra: Finding the Origins of Music in the World's Wild Places*. Profile Books, London.

Kulenkampff, J. 1981. "Music Considered as a Way of Worldmaking." *Journal of Aesthetics and Art Criticism* 39(3): 254–8.

Lakoff, G., and Johnson, M. 2003. *Metaphors We Live By*, 2nd ed. University of Chicago Press, Chicago.

Landy, L. 2006. "Electroacoustic Music Studies and Accepted Terminology: You Can't Have One without the Other." *Proceedings of the 3rd Electroacoustic Music Studies Network Conference (EMS06)*, Beijing, China.

Landy, L. 2007. *Understanding the Art of Sound Organization*. MIT Press, Cambridge, MA.

Lee, B. 1996. "Correspondence Analysis." *ViSta: The Visual Statistics System, UNC L.L. Thurstone Psychometric Laboratory Research Memorandum* 94–1(c): 63–78.

Lee, K. 2018. "Alice Shields: A Brief Introduction into Her Life and Work." *IUSB Graduate Research Journal* 5: 27–36.

Leman. M. 2008. *Embodied Music Cognition and Mediation Technology*. Press, Cambridge, MA.

Leman, M. 2012. "Musical Gestures and Embodied Cognition." *Actes des Journées d'Informatique Musicale* (JIM 2012), May 9–11, Mons, Belgium.

Lévi-Strauss, C. 1975. *The Raw and the Cooked: Introduction to a Science of Mythology I*. Translated by John Weightman and Doreen Weightman. Harper Colophon Books, New York.

Lim, Y., Donaldson, J., Kunz, B., Royer, D., Ramalingam, S., Thirumaran, S., and Stolterman, E. 2008. "Emotional Experience and Interaction Design." In Christian Peter and Russell Beale (eds.), *Affect and Emotion in Human-Computer Interaction: From Theory to Applications*: 116–29. Springer, Berlin.

Lin, C. "The Impact of the Development of the Fortepiano on the Repertoire Composed for It from 1760–1860." Doctoral dissertation, University of North Texas, Denton.

Luening, O. 1964. "Some Random Remarks about Electronic Music." *Journal of Music Theory* 8(1): 89–98.

Marcell, M. M., Borella D., Greene, M., Kerr, E., and Rogers, S. 2000. "Confrontation Naming of Environmental Sounds." *Journal of Clinical and Experimental Neuropsychology* 22(6): 830–64.

Marler, P. 2001. "Origins of Music and Speech: Insights from Animals." In Nils L. Wallin, Björn Merker, and Steven Brown (eds.), *The Origins of Music*: 31–48. MIT Press, Cambridge, MA.

Massumi, B. 2002. *Parables for the Virtual: Movement, Affect, Sensation*. Duke University Press, Durham, NC.

Massumi, B. 2010. "The Future Birth of the Affective Fact: The Political Ontology of Threat." In Melissa Gregg and Gregory J. Seigworth (eds.), *The Affect Theory Reader*: 52–70. Duke University Press, Durham, NC.

McCartney, A. S. J. 1999. "Sounding Places: Situated Conversations through the Soundscape Compositions of Hildegard Westerkamp." Doctoral dissertation, York University, Toronto, Canada.

McFarlane, M. 2001. "The Development of Acousmatics in Montréal." *eContact!* 6(2). Accessed July 5, 2019, at https://econtact.ca/6_2/mcfarlane_acousmatics.html.

Meelberg, V. 2006. "New Sounds, New Stories: Narrativity in Contemporary Music." Doctoral dissertation, Leiden University Press, Leiden, the Netherlands.

Meelberg, V. 2009. "Sonic Strokes and Musical Gestures: The Difference between Musical Affect and Musical Emotion." *Proceedings of the 7th Triennial Conference of European Society for the Cognitive Sciences of Music* (ESCOM 2009), Jyväskylä, Finland: 324–7.

Merer, A., Ystad, S., Kronland-Martinet, R., and Aramaki, M. 2007. "Semiotics of Sounds Evoking Motions: Categorization and Acoustic Features." In Richard Kronland-Martinet, Sølvi Ystad, and Kristoffer Jensen (eds.), *Computer Music Modeling and Retrieval: Sense of Sounds*: 139–58. Springer, Berlin.

Merker, B. 2012. "The Vocal Learning Constellation: Imitation, Ritual Culture, Encephalization." In Nicholas Bannan (ed.), *Music, Language, and Human Evolution*. Oxford University Press, Oxford, UK.

Meyer, L. B. 1956. *Emotion and Meaning in Music*. University of Chicago Press, Chicago.

Mikkonen, K. 2011. "'There Is No Such Thing as Pure Fiction': Impossible Worlds and the Principle of Minimal Departure Reconsidered." *Journal of Literary Semantics* 40(2): 111–31.

Miller, G. A. 2003. "The Cognitive Revolution: A Historical Perspective." *TRENDS in Cognitive Sciences* 7(3): 141–4.

Mithen, S. 2005. *The Singing Neanderthals: The Origins of Music, Language, Lind and Body*. Harvard University Press, Cambridge, MA.

Moore, B. C. J., and Hedwig, G. 2002. "Factors Influencing Sequential Stream Segregation." *Acta Acustica United with Acustica* 88(3): 320–33.

Moore, B. C. J., and Hedwig, G. 2012. "Properties of Auditory Stream Formation." *Philosophical Transaction of the Royal Society B* 367(1591): 919–31.

Murphy, T. S. 2005. "The Negation of a Negation Fixed in a Form: Luigi Nono and the Italian Counter-culture 1964–1979." *Cultural Studies Review* 11(2): 95–109.

Nagy, G. 1990. *Pindar's Homer: The Lyric Possession of an Epic Past*. Accessed August 8, 2020, at http://nrs.harvard.edu/urn-3:hul.ebook:CHS_Nagy.Pindars_Homer.1990.

Nattiez, J. 1990. *Music and Discourse*. Translated by Carolyn Abbate. Princeton University Press, Princeton, NJ.

Nilsson, P. A. 2018. "Notions of Experiment in Electroacoustic Music." *Proceedings of the 14th Electroacoustic Music Studies Network Conference* (EMS18), Florence, Italy: 108–10.

Nono, L. 1960. "The Historical Reality of Music Today." *The Score*, July: 41–5.

Nussbaum, C. O. 2007. *The Musical Representation: Meaning, Ontology, and Emotion*. MIT Press, Cambridge, MA.

Ohala, J. J. 1983. "Cross-language Use of Pitch: An Ethological View." *Phonetica* 40(1): 1–18.

Oliveros, P. 2005. *Deep Listening: A Composer's Sound Practice*. iUniverse, Bloomington, IN.

Olsen, D., and Nelson, M. J. 2017. "The Narrative Logic of Rube Goldberg Machines." *Proceedings of the International Conference on Interactive Digital Storytelling*: 104–16. Springer, Cham, Germany.

Omar, R., Henley, S. M. D., Bartlett, J. W., Hailstone, J. C., Gordon, E., Sauter, D. A., Frost, C., Scott, S. K., and Warren, J. D. 2011. "The Structural Neuroanatomy of Music Emotion Recognition: Evidence from Frontotemporal Lobar Degeneration." *Neuroimage* 56(3): 1814–21.

Oram, D. 1972. *An Individual Note: Of Music, Sound and Electronics*. Galliard Paperbacks, London.

Orgs, G., Lange, K., Dombrowski, J. H., and Heil, M. 2006. "Conceptual Priming for Environmental Sounds and Words: An ERP Study." *Brain and Cognition* 62(3): 267–72.

Oswald, J. 1985. "Plunderphonics, or Audio Piracy as a Compositional Prerogative." Lecture presented at Wired Society Electro-Acoustic Conference in Toronto. Accessed September 22, 2020, at http://www.plunderphonics.com/xhtml/xplunder.html.

Östersjö, S. 2008. "SHUT UP 'N' PLAY! Negotiating the Musical Work." Doctoral dissertation, Malmö Academies of Performing Arts, Sweden.

Oxford Dictionary of English, 3rd ed. Oxford University Press, 2012.

Özcan, E. 2008. "Product Sounds: Fundamentals & Applications." Doctoral dissertation, Industrial Design Department, Delft Technical University, the Netherlands.

Özcan, E., and van Egmond, R. 2007. "Memory for Product Sounds: The Effect of Sound and Label Type." *Acta Psychologica* 126(3): 196–215.

Özcan, E., and van Egmond, R. 2009. "The Effect of Visual Context on the Identification of Ambiguous Environmental Sounds." *Acta Psychologica* 131(2): 110–19.

Paivio, A. 1990. *Mental Representations: A Dual Coding Approach*. Oxford University Press, New York.

Parlak, E. 1998. "Türkiye'de El ile (Tezenesiz) Bağlama Çalma Geleneği ve Çalış Teknikleri." Doctoral dissertation, İstanbul Teknik Üniversitesi, Turkey.

Partridge, C. 2007. "King Tubby Meets the Upsetter at the Grass Roots of Dub: Some Thoughts on the Early History and Influence of Dub Reggae." *Popular Music History* 2(3): 309–31.

Patel, A. D. 2003. "Language, Music, Syntax and the Brain." *Nature Neuroscience* 6(7): 674–81.

Patel, A. D. 2007. *Music, Language, and the Brain*. Oxford University Press, New York.

Paul, D. 1997. "Karlheinz Stockhausen by David Paul." Originally published in *Seconds* 44, 1997. Accessed February 17, 2014, at https://web.archive.org/web/20140404172943/http://www.stockhausen.org/stockhausen%20_by_david_paul.html.

Pecher, D., Zeelenberg, R., and Barsalou, L.W. 2003. "Verifying Different Modality Properties for Concepts Produces Switching Costs." *Psychological Science* 14: 119–24.

Peignot, J. 1960. "De la Musique Concrète à L'acousmatique." *Esprit* 280(1): 111–20.

Peirce, C. S. 1885. "On the Algebra of Logic: A Contribution to the Philosophy of Notation." *American Journal of Mathematics* 7(2): 180–96.

Pinch, T., and Trocco, F. 1998. "The Social Construction of the Early Electronic Music Synthesizer." *Icon* 4: 9–31.

Pinker, S. 2003. *The Blank Slate: The Modern Denial of Human Nature*. Penguin Books, London.

Plato. [*c.* 380 BC] 1985. *The Republic*. Translated by Richard W. Sterling and William C. Scott. Norton, New York.

Prieto, E. 2020. "Nelson Goodman: An Analytic Approach to Music and Literature Studies." In Delia da Sousa Correa (ed.), *The Edinburgh Companion to Literature and Music*. Edinburgh University Press, Edinburgh, Scotland.

Radigue, E. 2009. "The Mysterious Power of the Infinitesimal." *Leonardo Music Journal* 18: 47–9.

Raimbault, M., and Dubois, D. 2005. "Urban Soundscapes: Experiences and Knowledge." *Cities* 22(5): 339–50.

Reich, S. 1987. *Early Works*. Elektra Nonesuch, New York, . Liner notes.

Richman, B. 2001. "How Music Fixed 'Nonsense' into Significant Formulas: On Rhythm, Repetition, and Meaning." In Nils L. Wallin, Björn Merker, and Steven Brown (eds.), *The Origins of Music*: 301–14. MIT Press, Cambridge, MA.

Roads, C. 2001. *Microsound*. MIT Press, Cambridge, MA.

Roads, C. 2009. "Touche pas." Program notes. CREATE Concert, November 13, Lotte Lehmann Concert Hall, University of California, Santa Barbara, CA.

Roads, C. 2015. *Composing Electronic Music: A New Aesthetic*. Oxford University Press, New York.

Roads, C. 2016. Personal communication on March 7, 2016.

Robindoré, B. 2005. "Forays into Uncharted Territories: An Interview with Curtis Roads." *Computer Music Journal* 29(1): 11–20.

Rodgers, T. 2010. *Pink Noises: Women on Electronic Music and Sound*. Duke University Press, Durham, NC.

Rose, T. 1994. *Black Noise: Rap Music and Black Culture in Contemporary America*. Wesleyan University Press, Hanover, NH.

Rosenbloom, E. 2011. *Morton Subotnick on the Creation and Legacy of Silver Apples of the Moon*. Interview accessed November 16, 2020, at https://www.ascap.com/news-events/articles/2011/04/p-Morton-Subotnick-on-the-Creation-and-Legacy-of-Silver-Apples-of-the-Moon.

Russolo, L. 1967. "The Art of Noise: Futurist Manifesto, 1913." In *A Great Bear Pamphlet*. Something Else Press, New York.

Ryan, M. 1980. "Fiction, Non-factuals, and the Principle of Minimal Departure." *Poetics* 9(4): 403–22.

Salimpoor, V. N., van den Bosch, I., Kovacevic, N., McIntosh, A. R., Dagher, A., and Zatorre, R. J. 2013. "Interactions between the Nucleus Accumbens and Auditory Cortices Predict Music Reward Value." *Science* 340: 216–19.

Samson, J. 1977. *Music in Transition: A Study of Tonal Expansion and Atonality, 1900–1920*. Dent, London.

Savage, P. E., Brown, S., Sakai, E., and Currie, T. E. 2015. "Statistical Universals Reveal the Structures and Functions of Human Music." *Proceedings of the National Academy of Sciences* 112(29): 8987–92.

Scherer, K. R., and Oshinsky, J. S. 1977. "Cue Utilization in Emotion Attribution from Auditory Stimuli." *Motivation and Emotion* 1(4): 331–46.

Schmidt, J. R. 1981. "Expansion of Sound Resources in France, 1913–1940, and Its Relationship to Electronic Music." Doctoral dissertation, University of Michigan, Ann Arbor.

Schutz, A. 1967. *The Phenomenology of the Social World*. Translated by George Walsh and Frederick Lehnert. Northwestern University Press, Evanston, IL.

Schwartz, D. A., Weaver, M., and Kaplan, S. 1999. "A Little Mechanism Can Go a Long Way." Open peer commentary on Barsalou's "Perceptual Symbol Systems" in *Behavioral and Brain Sciences* 22(4): 631–2.

Seigworth, G. J., and Gregg, M. 2010. "An Inventory of Shimmers." In Melissa Gregg and Gregory J. Seigworth (eds.), *The Affect Theory Reader*: 1–28. Duke University Press, Durham, NC.

Shouse, E. 2005. "Feeling, Emotion, Affect." *M/C Journal* 8(6). Accessed June 6, 2014, at http://journal.media-culture.org.au/0512/03-shouse.php.

Sievers, B., Polansky, L., Casey, M., and Wheatley, T. 2013. "Music and Movement Share a Dynamic Structure That Supports Universal Expressions of Emotion." *PNAS* 110(1): 70–5.

Simoni, M. (ed.). 2006. *Analytical Methods of Electroacoustic Music*. Routledge, New York.

Simoni, M. 2018. "The Audience Reception of Algorithmic Music." In Roger T. Dean and Alex McLean (eds.), *The Oxford Handbook of Algorithmic Music*: 531–57. Oxford University Press, New York.

Sloboda, J. A. 2005. *Exploring the Musical Mind: Cognition, Emotion, Ability, Function*. Oxford University Press, Oxford, UK.

Smalley, D. 1991. "Acousmatic Music—Does It Exist?" In A. Vande Gorne (ed.), *Vous Avez Dit Acousmatique? Lien, Revue d'Esthétique Musicale*: 21–2. Musiques et Recherches, Ohain, Belgium.

Smalley, D. 1996. "The Listening Imagination: Listening in the Electroacoustic Era." *Contemporary Music Review* 13(2): 77–107.

Smalley, D. 1997. "Spectromorphology: Explaining Sound-Shapes." *Organised Sound* 2(2): 107–26.

Sousa, D. P. E. 2008. *Nuria Schöenberg-Nono: Around Music*. Interview accessed May 19, 2014, at https://artenotempo.pt/en/nuria-2/.

Spiegel, L. 1992. "An Alternative to a Standard Taxonomy for Electronic and Computer Instruments." *Computer Music Journal* 16(3): 5–6.

Spinoza, B. 1994. *A Spinoza Reader: The Ethics and Other Works*. Edited and translated by Edwin M. Curley. Princeton University Press, Princeton, NJ.

Stienstra, J. 2015. "Embodying Phenomenology in Interaction Design Research." *Interactions* 22(1): 20–1.

Stockhausen, K. 1962. "The Concept of Unity in Electronic Music." Translated by Elaine Barkin. *Perspectives of New Music* 1(1), Autumn: 39–48.

Stockhausen, K. 1972. *Four Criteria of Electronic Music*. Lecture given at Oxford Union, Oxford, UK, on May 6. Video available from http://www.karlheinzstockhausen.org.

Stockhausen, K. 1989. "Four Criteria of Electronic Music." In Robin Maconie (ed.), *Stockhausen on Music: Lectures and Interviews*: 88–111. Marion Boyars, London.

Stowell, D., and Plumbley, M. D. 2014. "Large-Scale Analysis of Frequency Modulation in Birdsong Data Bases." *Methods in Ecology and Evolution* 5(9): 901–12.

Tajadura-Jiménez, A., and Västfjäll, D. 2008. "Auditory-Induced Emotion: A Neglected Channel for Communication in Human-Computer Interaction." C. Peter and R. Beale (eds.), *Affect and Emotion in HCI*, LNCS 4868: 63–74. Springer, Berlin.

Taruffi, L., and Küssner, M. B. 2019. "A Review of Music-Evoked Visual Mental Imagery: Conceptual Issues, Relation to Emotion, and Functional Outcome." *Psychomusicology: Music, Mind, and Brain*: 29(2–3): 62–74.

Thagard, P. 2014. "Cognitive Science." In Edward N. Zalta (ed.), *The Stanford Encyclopedia of Philosophy*. Accessed June 12, 2014, at http://plato.stanford.edu/archives/fall2014/entries/cognitive-science/.

Thomas, N. J. T. 2010. "Mental Imagery." In Edward N. Zalta (ed.), *The Stanford Encyclopedia of Philosophy*. Accessed May 20, 2014, at http://plato.stanford.edu/archives/spr2014/entries/mental-imagery/.

Toop, R. 1981. "Stockhausen's Electronic Works: Sketches and Work-Sheets from 1952–1967." *Interface Journal of New Music Research* 10(3–4): 149–97.

Trainor, L. J., Tsang, C. D., and Cheung, V.H. 2002. "Preference for Sensory Consonance in 2- and 4-Month-Old Infants." *Music Perception* 20(2): 187–94.

Trehub, S. 2001. "Human Processing Predispositions and Musical Universals." In Nils L. Wallin, Björn Merker, and Steven Brown (eds.), *The Origins of Music*: 427–48. MIT Press, Cambridge, MA.

Truax, B. 1984. *Acoustic Communication*, 2nd ed. Ablex, Norwood, NJ.

Truax, B. 1996. "Soundscape, Acoustic Communication and Environmental Sound Composition." *Contemporary Music Review* 15(1): 49–65.

Tucker, D. G. 1976. "The Invention and Early Use of the Telephone." *IETE Journal of Research* 22(3): 101–6.

Vaggione, H. 2001. "Some Ontological Remarks about Music Composition Processes." *Computer Music Journal* 25(1), March: 54–61.

VanDerveer, N. J. 1979. "Ecological Acoustics: Human Perception of Environmental Sounds." Doctoral dissertation, University of Cornell, Ithaca, NY.

Van Petten, C., and Rheinfelder, H. 1995. "Conceptual Relationships between Spoken Words and Environmental Sounds: Event-Related Brain Potential Measures." *Neuropsychologia* 33(4): 485–508.

Varèse, E., and Wen-chung, C. 1966. "The Liberation of Sound." *Perspectives on New Music* 5(1): 11–19.

Vigliocco, G., Vinson, D. P., Damian, M. F., and Levelt, W. 2002. "Semantic Distance Effects on Object and Action Naming." *Cognition* 85(3): B61–B69.

Vines, B. W., Krumhansl, C. L., Wanderley, M. M., Dalca, I. M. and Levitin, D. J. 2011. "Music to My Eyes: Cross-modal Interactions in the Perception of Emotions in Musical Performance." *Cognition* 118(2): 157–70.

Walton, K. 1994. "Listening with Imagination: Is Music Representational?" *Journal of Aesthetic and Art Criticism* 52(1): 47–61.

Warren Jr., W. H., Kim, E. E., and Husney, R. 1987. "The Way the Ball Bounces: Visual and Auditory Perception of Elasticity and Control of the Bounce Pass." *Perception* 16(3): 309–36.

Weale, R. 2006. "Discovering How Accessible Electroacoustic Music Can Be: The Intention/Reception Project." *Organised Sound* 11(2): 189–200.

Wetherell, M. 2012. *Affect and Emotion: A New Social Science Understanding*. Sage, London.

Whittall, A. 2011. "Electroacoustic Music." In Alison Latham (ed.), *The Oxford Companion to Music*. Oxford University Press. Accessed September 2, 2020, at https://www.oxfordmusiconline.com/page/the-oxford-companion-to-music.

Windsor, L. 1995. "A Perceptual Approach to the Description and Analysis of Acousmatic Music." Doctoral dissertation, Department of Music, University of Sheffield, UK.

Windsor, L. 2000. "Through and around the Acousmatic: The Interpretation of Electroacoustic Sounds." In Simon Emmerson (ed.), *Music, Electronic Media and Culture*: 7–35. Ashgate, Burlington, VT.

Winter, T. 2015. "Delia Derbyshire: Sound and Music for the BBC Radiophonic Workshop, 1962–1973." Doctoral dissertation, Department of Music, University of York, UK.

Wishart, T. 1986. "Sound Symbols and Landscapes." In Simon Emmerson (ed.), *The Language of Electroacoustic Music*: 41–60. Macmillan Press, Houndmills, UK.

Wishart, T. 1996. *On Sonic Art*. Simon Emmerson (ed.). Harwood Academic, Amsterdam.

Wurmfeld, S. 1993. "Presentational Painting." In *Catalogue for the Art Gallery, Hunter College, MFA Building*, October 20–November 20.

Xenakis, I. 1992. *Formalized Music: Thought and Mathematics in Composition*, rev. ed. Additional material compiled and edited by Sharon Kanach. Pendragon Press, Stuyvesant, NY.

Yost, W. A. 2015. "Psychoacoustics: A Brief Historical Overview." *Acoustics Today* 11(3): 46–53.

Zampronha, E. 2005. "Gesture in Contemporary Music: On the Edge between Sound Materiality and Signification." In *Transcultural Music Review* 9: n.p.

Zoble, E. J., and Lehman, R. S. 1969. "Interaction of Subject and Experimenter Expectancy Effects in a Tone Length Discrimination." *Behavioral Science* 14(5): 357–63.

Index

abstract causality 113
abstract syntax 46
abstraction 3, 4, 18, 68–70, 159, 160, 162–8
   electronic gesture and 102, 112–13
   musical behavior and 30, 32, 45–7, 50, 51, 53
   physical domain and 144–6, 149
   semantic domain and 150, 153, 155
   worldmaking and 122, 125, 127, 130, 137
abstractness 3, 123, 137–8, 169, 172
   as negation of reality 2, 50–1, 128
accompanying gestures 107
acousmatic music 16–17, 46, 50–1
acoustic 8–9, 17, 33, 37, 45, 50, 74, 100–3, 112, 115, 137, 147, 154
   communication 103
   ecology 148, 154
   environment 103, 106, 148, 154
   information 160, 164
   properties 26, 28, 43, 53, 66, 136, 144, 159, 163
   wave 8
action/perception feedback loop 38–9, 49, 51
action descriptors 90, 111
Adorno, T. W. 30 n.1, 49, 50
Aesthetic Theory (Adorno) 49
affect 30, 44, 106, 121, 126, 147
   diegetic affordances and 131–7
   in music 33–4
affective descriptors 90, 91, 112, 144
affordances 3, 18, 28, 39, 42, 51, 121, 142, 172–3
   diegetic, and affect 131–7
   model of 104–6
Alexenberg, M. 127
algorithmic
   melodies 19
   music 60–1, 122
   processes 117
Amacher, M. 38
amalgamation of musical languages 47–8

ambient noise 53, 152
ambient texture 19, 73, 74, 116
amorphous sequences 53
Animal Collective 19
anthrophony 24
anticipation (mechanisms) 34, 44, 45, 103, 110, 114, 126, 129–30
appraisal 52
   affective 23, 30, 34, 45, 90, 137, 145, 148
   emotion 31
   music 23, 31, 33, 34, 56, 100, 148, 149
appraisal descriptors 91, 139
ARP 2500 synthesizer 18, 43
artistic material 49–52, 124
art music 17
"art of noises" 10
audio decorrelation 69, 71, 116
audiovisual 60, 64
auditory descriptors 91, 144, 150
auditory perception 1, 2, 36, 50, 61, 79, 86, 89

Back to the Future (film) 76
Bal, M. 129
Baldwin effect 26
Ballas, J. A. 100, 150
band-pass filters 67
Bar, M. 151
Barbieri, C. 19
Barrett, N. 160, 161, 163–6
Barsalou, L. 102
Basanta, A. 60
Bayle, F. 17
BBC Radiophonic Workshop 19
Beatles 138
Bebe and Louis Barron 20
Beethoven, L. 76
Berio, L. 41, 138
Bhatara, A. 147–8
biomusicology 24–6, 30, 56
biophonic score 28
biophony 24, 28

Birdfish (Çamcı) (composition) 62, 65, 81, 110–11, 118, 135, 144–7, 155–9, 163
　abstract leitmotif in 137–8
　contextual cues in 150–1
　fabula and 152
　form of 67–9
　in drawing, participant's impression of *132*
　narrative in 130–1
　reverberation in 134
　science fiction in 141
　sound design of 65–7
bird vocalizations 66
bistability 146
Boulez, P. 1, 40, 42
brain stem reflex 31, 33, 34
Bridger, M. 58
Briscoe, D. 19
Brown, S. 26
Bruckner, A. 138
Buchla, D. 18
Buchla synthesizer 18

Cage, J. 9, 11
Caramazza, A. 154
Carlos, W. 19, 20
causal listening 45
causal network 5, 77, 112–15, 118–19
Chadabe, J. 10 n.1
Chapin, H. 147
Cheung, V. H. 26
Chion, M. 45
Chowning, J. 15
Christmas 2013 (Çamcı) (composition) 62, 71–2, 76, 80–1, 134, 141, 148, 157–9
　electronic gestures in 112, 114–15
　form of 73–4
　musical quotation in 139–41
　sound design of 72–3
Ciani, S. 18, 67
Circling Blue (Hartman) (composition) 47
Clockwork Orange, A (Kubrick) (film) 21
cognition/cognitive 3–4, 31, 34–5, 50, 52, 55–62, 99–102, 105–7, 123, 133, 152, 172, 174
　auditory 56, 143, 171
　psychology 3, 55, 102
cognitive representations 52–3, 100

cognitive revolution 55–6
Coker, W. 107
Cologne studio 12, 19, 40–2, 46, 67 n.2
Come Out (Steve Reich) (composition) 20
Comme une symphonie envoi a Jules Verne (Henry) (composition) 138
communication model 103
comparative analysis 92–3
composer, as listener 35–40, 49
composer's instincts 13, 23, 42, 43
composer's intentions and listener's experience, relationship between 59–60
composer's presence, in work 159–60
composition. *See also individual compositions* 2, 13–16, 18, 23, 29, 35–7, 39–42, 49, 51–2, 55, 59, 108, 172
composition and decomposition 122
computer music 15–16
Computer Music Journal 160
concept descriptors 91, 144, 150
concert grand 15
concrete causality 113
concrete hearing 35, 38–9
connotations 32, 43–5, 52–3, 123, 125, 138, 153, 163
connotative symbols 32
context frames 151
contextual cues 150–1
Continuous Variation (Deleuze) 132
conventions, musical 30
correspondence analysis 93, *94*
creative constraints 52

Dack, J. 43
Daft Punk 67
Debussy, C. 138
deejaying 20
deformation in worldmaking 122
Delage, B. 154
Delalande, F. 58–9, 61, 107
delay (audio effect) 20, 65, 69, 72, 73, 116
delay lines 75–6
deletion and supplementation in worldmaking 122
Deleuze, G. 33, 127–8, 132
demand characteristics 80
Demers, J. 43, 44, 49–50, 171
Derbyshire, D. 20

descriptor categorization 3, 81, 89–92, 94, 102, *111*, 112, 139, 159
designative meaning 110, 127
Deutsch, D. 149
Diegese (Çamcı) (composition) 62, 112, 113–17, 141, 147, 156, 159
  affordances and affect in 134, 135–6
  form of 77–8, *77*
  musical quotations in 138, 141
  sound design of 75–7
diegesis 3, 75, 119, 122–3, 150, 160, 167–9.
  See also diegetic actor
  affordances and affect and 131–7
  coalescence of mimesis and 125–6
  interdisciplinary contextualization of 123–5
  narrativity and 127–31
  physical and semantic domains interaction and 155–7
  presentationality and 126–7
  as spatiotemporal universe 123, 125
"Diegesis as a Semantic Paradigm for Electronic Music" (Çamcı) 74–5
diegetic actor 75, 150, 163, 167
  as music 141–2
  music as 137–41
diegetic affordances 5, 121, 134, 172
diegetic sound 75, 125, 141, 164
digital audio workstation (DAW) 39, 51, 75, 115
discourse analysis 61, 93–4, *94–7*
dishabituation 148
Doctor Who (TV series) 20
Doornbusch, P. 10 n.1
Doppler shifts 167
dramaturgical model, of sonic behavior 113
Dubois, D. 53, 110, 154
Duchamp, M. 118
Dudley, H. 67 n.2
dynamic/dynamics 1, 25, 28, 40, 79, 106, 146, 148, 153, 174
  contour 66, 100
  properties 29, 144
  qualities 147
  visualization 87–8

earwitness accounts 103
ecological approach and acoustic communication 103

Ecological Approach to Visual Perception, The (Gibson) 104–5
ecological perception 104
eContact (journal) 75
Edison, T. 8
effective gestures 107
Eimert, H. 12, 40–2, 57
Eitan, Z. 25
elasticity, of sound material 39
El-Dabh, H. 9
electroacoustic music 17, 58–60
electronic gesture 99
  and events in environment 99–100, 106
  environmental sounds and 100–2
  mental representation models and 102–6
  significance of 106–9; (see also gesture)
electronic medium 1–2, 8–9, 13–14, 18, 20, 35, 38–43, 45, 49–52, 63, 109, 115, 138, 171
electronic music 14–18. *See also individual entries*
  cognitive continuum of 3, 45
  electronic medium and 8–9, 13–14
  experiential idiosyncrasies of 4, 34–53
  physical domain of 144–50, 155–7
  science fiction and 141–2
  semantic domain of 150–5
  studio, birth of 11–13
Electronic Studies (Stockhausen) (composition) 46
elektronische Musik 12, 13, 46
Element Yon (Çamcı) (composition) 62, 69, 113, 114, 117, 141, 144–6, 149, 154, 158, 159, 163
  diegetic affordances in 134–5
  form of 70–1
  sound design of 70
embodied engagement 34, 156, 157, 167
embodied experience 130, 145
embodied meaning 110, 125, 127
embodied presence 1, 145, 159
Emmerson, S. 4, 46
emotion 24–5, 34, 60, 108, 112, 114, 123, 130, 133, 141, 147, 154, 155, 174
  and affect, comparison of 34
  music and 31–3
emotional contagion 31, 32

emotion descriptors 91, 144
emphatic listening 58
energy transfer model 103
envelope generator 66, 70
environmental sounds 53, 62, 99–103, 106, 110, 115, 125–6, 149, 152, 154, 171
    categorization of 111–12
    as intentional gestures 117 n.2
    and musical sound compared 100–1
    organic sounds and 163
episodic memory 31–2
Epitaph for Aikichi Kuboyama (Eimert) (composition) 57
esthesis/esthesic 48–50, 52–4, 104, 107, 112, 119, 123, 128, 129, 169
    intentionality and 117, 118
    worldmaking and 123
Ethics (Spinoza) 33
evaluative conditioning 32
events, in environment 99–100
    environmental sounds and 100–2
    mental representation models and 102–6
experienced listeners 158–9
experiment bias 80
Expression of Zaar, The (El-Dabh) (composition) 9
extradiegetic narrator 159

fabula 129, 131, 152
featural descriptors 91
Ferrari, L. 46
Field, A. 43, 45
figurative gestures 108
figurative listening 58
figure and ground 147, 150, 152, 153
film 3, 20, 37, 76, 115, 121, 133, 156, 172
film sound 20, 75, 124, 125, 141, 142
filter/filtering 12, 37, 67, 70, 72–4, 147, 152, 153, 160–2, 164–6, 168
fixed music 64–5
Flicker Tone Pulse (Roads) (album) 64 n.1
Forbidden Planet (film) 20
Four Criteria of Electronic Music" (Stockhausen) 27
frequency 26, 73, 75, 103, 116, 135, 137, 146, 147, 152, 154, 160
    domain 39, 51, 72
    distribution 92, 93
    impulses 161, 168
    range 70, 71, 74
    rumbles 68, 134, 157
    spectrum 28, 67, 71, 100
    textures 116, 165
frequency modulation (FM) synthesis 65, 66, 70
frisson 30, 148
Fritz, T. 32
fundamental frequency 135, 137, 154
fundamental noises 11
future nostalgia 72, 140

Gaver, W. 103, 135
general impressions 79–83, 86, 89, 92–7, 112, 114, 117, 118, 145, 149–52, 155, 158. *See also* diegesis
generative 62, 63, 78, 117, 122, 124
Genette, G. 123, 125, 129, 157, 160
geophony 24
Gesang der Jünglinge (Stockhausen) (composition) 42
gestalt 63, 152–4
gestalt perception 107, 108
    of gesture 117
gesture 2, 3, 5, 25, 33, 36, 106–9, 156–60, 166–9
    coexisting, in temporal and spatial configurations 115–17
    habituation and 148–9
    implying intentionality 117–19
    listening and 68–70, 72–4, 77–9, 88
    in *Little Animals* 162–5
    as meaningful narrative unit 109–12
    operating within causal network 112–15
    semantic domain and 152–4
    stream segregation and 147
    worldmaking and 121, 123, 126, 135–8
Gibson, J. 104, 109, 112, 126, 131, 133
Gillespie, S. 30 n.1
Goldstein, B. 105
Goodman, N. 122, 123
Good Vibrations (The Beach Boys) (song) 20 n.2
Grammophonmusik 9
Grandmaster Flash 20
Granot, R. 25

grain (sound) 76, 113, 135, 136, 147
granular synthesis 19, 65, 66, 76, 135, 136, 147
graphic score 163
Great Animal Orchestra, The (Krause) 24, 28
Gritten, A. 107, 117
Groupe de Recherches Musicale 12
Guastavino, C. 53, 110, 154
Guattari, F. 33, 127–8, 132
Gygi, B. 111

habituation 147–50
habituation syndrome 148
Haertman, H. 47
harmonic 28, 30, 67, 70, 73, 163
   patterns 100
   progression 113, 159
   resolution 71, 74, 113
Hatten, R. 107, 108
Hajime (Çamcı) (composition) 76
Hedwig, G. 146
Henry, P. 58, 138
Herman, D. 129
Hill, A. 60
Hindemith, P. 9
Hitchcock, A. 20
Howard Jr., J. H. 100
Howe, H. 18
human-centred research 57
human gestures 107
Huron, D. 30, 34, 45, 106, 148
Hymnen (Stockhausen) (composition) 138

Iazzetta, F. 107, 108
icons (Charles Sanders Peirce) 101
Ihde, D. 35
image processes 44
Imaginary Landscape No. 1 (Cage) (composition) 9
indexes (Charles Sanders Peirce) 101
Individual Note of Music, Sound and Electronics, An (Oram) 27
inner hearing 35–40, 51, 108
inscription 36, 163
Insectarium (*Birdfish*) (Çamcı) (installation) 65
instinctual knowledge 106

instrumental music 29–30, 37, 44, 57
   communication of expression in 108
   motormimetic component in 108–9
   narrative in 130
   neutral state and 153
   presentationality in 127
instrumental musique concrète 47
instrumental tracks 20
intentionality 2, 5, 49, 99–101, 107, 123, 172
   gesture and 117–19, 121
Intention Reception Project 59–60
interface
   keyboard 19
   musical 18
   software 82–5
intonarumori. *See* noise intoners
intonation 141, 154
invariants 37, 104–6
I of IV (Pauline Oliveros) (composition) 20
It's Gonna Rain (Steve Reich) (composition) 19
Ives, C. 138

Jackson, M. 138
Jamaican sound system 20
Jekosch, U. 117
Johnson, M. 99, 115
Juslin, P. 31, 143

Kaplan, S. 106
Karbusický, V. 27, 47, 57, 142
Karlheinz Stockhausen's Kontakte and Narrativity" (Dack) 43
keyboard (piano) 18–19
Kidd, G. R. 111
kinesthesis (aural, visual) 25, 133–4
King, E. 107, 117
King Tubby 20
Koelsch, S 29, 32
Kontakte (Stockhausen) (composition) 42, 71
Kool Herc 20
kopuz 28
Korg MS-20 synthesizer 70
Kraftwerk 67
Krause, B. 24, 28

Kubrick, S. 21

Lachenmann, H. 47
La Fabbrica Illuminata (Nono) (composition) 155, 156
Lakoff, G. 99, 115
Landy, L. 17, 59, 60
language grid 46
language of music 2, 27–30, 44, 47
leitmotif 65, 68, 69, 137–8
Let the World End (Birdfish) (Çamcı) (composition) 65
Levi-Strauss, C. 29, 127
listener's self-image, imprint of 61
listening, complexity of 43–6
listening behaviors 58–9
listening imagination 55, 62–4
Listening through the Noise (Demers) 171
Little Animals (Barrett) (composition) 160–1
   diegetic disposition of listener and 166–9
   gestural layers in 162–3
      organic and environmental sounds 163
      physical ausalities 164
      pitched and droning elements 164–5
   macroscale analysis of 161–2
   temporal flow in 165–6
live performances 65, 70
location and semantic gestalt 153
location descriptors 91, 150, 163
loudness 25, 30–2, 40, 103, 106, 146, 148, 153, 155, 168
low-frequency oscillation (LFO) 66, 67, 70
low-frequency pulses 68

McCartney, A. 59, 61
Maderna, B. 41
magnetic recording 9
Marcell, M. M. 101–2, 111
Martinville, E. 8
Massumi, B. 33, 131, 133
material
   of music 1, 2, 18, 21, 27–30, 32–5, 38, 40, 42, 44–7, 67, 69, 72, 74, 115, 127, 134, 137, 144
   sonic 39, 44, 69, 114, 123, 160, 167, 169
   of sound 18, 39, 51, 69, 72, 74, 78, 117, 142, 150, 162, 165

Matthews, M. 15
Meelberg, V. 33, 129
memory and experience, factors related to 27
mental gestures 108
mental imagery 31, 102, 112, 117
mental representation models 102–6
Merker, B. 26
meta descriptors 91, 158, 159
metaphoric primitives 44, 45
Meyer, L. B. 32–3, 44, 52, 153
Meyer-Eppler, W. 67 n.2
microfigures 79
micromontaging 66, 72, 78, 162
Miller, G. 56
mimesis, Platonic 124
mimetic discourse 46
minimal departure principle 128
modular synthesizer 19
Molino, J. 48, 117
moment form 71
Monk, T. 61
monophonic signal 69
Moog, R. 18
Moog Slim Phatty synthesizer 75
Moog synthesizer 19, 76
Moore, B. C. J. 146
motormimetic component 108–9
movement and music, relationship between 25
   gesture and 107
Mullins, T. 150
multichannel audio 155
multi-layered spatiality 116
multimodality 99–100, 103, 110
multiple-timeline visualization 88
musical behavior 4, 8, 23, 56, 101
   foundations of 23–34
   cultural and biological factors 4, 23, 25–8, 30–2, 34, 44, 45, 47, 53, 107, 143
musical descriptors 90–1
musical expectancy 26, 32, 34, 44–5, 130
musical form 1, 76, 112, 137, 138, 141, 149
musical language. See language of music
musical meaning 1, 110, 112
   absolute 32
   referential 32
musical material. See material

musical sound and environmental sound, comparison of 100–1
musical source descriptors 139–41, 144, 163
musical universals 25–6, 33
musical vocabulary. *See* vocabulary of music
Music for the Funeral of Queen Mary (Purcell) (composition) 21
music psychology 56
musique concrète 12, 19, 41, 43, 46, 57, 122
  gesture and 110
  instrumental 47

narrative 36, 63, 68, 72, 74, 75, 117, 125, 126, 130–1, 134, 137, 143, 145, 146, 152, 161, 168, 169
  context 59, 124, 160
  construction 3, 5, 124, 150
  definition of 127–9
  framework 121, 123
  musical 106, 113, 129, 157
  structure 69, 114, 119, 128, 129
  as temporal 157–8
  unit 2, 3, 109–12, 118, 119
narrativity 5, 127–31, 172
narratology 3, 121, 123, 129
Nattiez, J. 48, 118
Nautik (*Birdfish*) (Çamcı) (composition) 65
negation of reality 2, 50–1, 128
neutral state 153
neuroscience 26–7, 29, 56, 143, 151, 154
New York Morning Telegraph 10
Nilsson, P. A. 17
noise 10–11, 53, 64, 115, 138, 148, 163. *See also* ambient noise
  generator 12, 37
  signal 67, 160
  source 67
noise intoners 11
non-diegetic sound 75, 124, 163, 167, 168
nonmusical sound 9–11, 44, 47, 158
Nono, L. 41, 155–6
notation 1, 30, 36, 38

object descriptors 90, 91, 111
object source descriptors 144

observer expectancy effect 80
Ohala, J. 136
Oliveros, P. 20
ondes Martenot 8
onomatopoeia category 91
Oram, D. 1, 13, 19, 27
ordering in worldmaking 122, 129
organized sound 11, 27, 60
Orgs, G. 61
orienting response 148
Oswald, J. 138
oscillator 12, 19, 37, 41, 46, 66–7
Özcan, E. 53

Painting With (Animal Collective) (album) 19
Paivio, A. 105
panning 36, 153, 161, 166
panorama (stereo) 73, 76, 116
Paris studio 12, 41, 43, 50
Parsifal (Wagner) (composition) 138
Patel, A. 29
Patterns of Consciousness (Caterina Barbieri) (album) 19
Peignot, J. 16
Peirce, C. S. 101
percecptual knowing 105–6
perceptual descriptors 90, 91, 111, 112
perceptual symbols 102, 105, 151
Perry, L. 20
personal computers 39–40
phonautograph 8
phonograph 8, 9, 11, 12
physical self, awareness of 127, 144–6, 159, 174
Piaget, J. 102, 117
piano 14, 29, 76–8, 112, 139–41
Piano Sonata No. 27 in E minor (Opus 90) (composition) 76
Pink Floyd 139 n.2
pitch 12, 26, 66–8, 73–4, 106, 108, 116, 154–5, 157–8, 163–5, 168
  gliding 66, 114, 141, 154, 163
  patterns 24
  qualities 25
  sequences 40
pitch shifting 72, 73, 113, 148, 162
plasticization, of music 12, 46

Plato 123–4
Plunderphonic (Oswald) (album) 138
plunderphonics 138
poiesis/poietic 48–52, 103, 104, 107, 112, 119, 121, 140, 159, 160, 169, 172
   intentionality and 117–18
   worldmaking and 123
polyphony 24, 150
post-Schaefferian era 43, 57
presentational acting 127
presentational object 145
presentational art 127
Presque Rien (Ferrari) (composition) 46
prototypical contexts 151
psychoacoustics 56, 167
psychology 100, 105, 107
   cognitive 3, 55, 56, 102
   of music 4, 26, 29, 31, 53, 56, 104
pulsar synthesis 65
Purcell, H. 21

quadraphonic sound 42
quality descriptors 91, 139, 140
quotation (music) 73, 75–8, 121, 138–41

Radigue, E. 18, 43
Radiodiffusion Française 12
radiophonic art 41
Raimbault, M. 110, 154
Ravel, J. M. 138
reality 1–3, 10, 14, 49–51, 53, 56, 60, 67, 102, 112, 121, 123, 125, 127, 128, 140, 143, 144, 162–3, 166, 167. *See also* virtual reality
   physical 3, 50, 56, 133, 146, 157, 158
real-time descriptors 79, 81–9, 92–4
reception theory 60
referential sound 37–8
Reich, S. 19–20
Reis, P. 8
representational acting 127, 145
representationality 68–70, 124, 125, 130, 137, 144–6, 155, 159, 160, 162–3, 167, 169
   abstractness and 3–4, 45, 50, 70, 112, 123
rests, significance of 68, 70
reverberation 68–70, 134, 167

reversed reverb tail 77, 78, 114
rhythmic grid 115, 148, 149, 166
rhythmic patterns, structures 9, 30, 149, 166, 168
Roads, C. 18, 40, 62, 75, 113, 114, 129, 135 n.1
Rosenbloom, E. 14, 38
Rózsa, M. 20
Rube Goldberg Machine 76, 115
Russolo, L. 9–11

Salimpoor, V. N. 27
sampling 19–20, 138, 149
Schaeffer, P. 12, 16, 44, 57
Schafer, M. 154
schemas 31, 45, 48–9, 102–3, 105, 117, 151
Schönberg-Nono, N. 156
Schwartz, D. A. 106
science fiction 20, 76, 141–2
selective auditory attention 153
self-representation sounds 154
semantic association 32, 53, 61, 134, 136
semantic context 150–2, 165
semantic gestalts 153–4
semantic processing 34, 44, 54, 144
semantics 3, 29, 34, 57, 60–2, 92–4, 110, 112, 125, 126, 128, 137, 138, 141, 143, 150–60, 163–5, 167–9
   effects of 150–2
semiology 4
semiotic model 48–9
serial music 12–13, 40–2, 46, 108
Shadowbands (Çamcı) (composition) 76
Shelton, J. R. 154
Shields, A. 51–2
"Shine On You Crazy Diamond" (Pink Floyd) (Song) 139 n.2
Sievers, B. 25
signpost gestures 163
Silent Night (Christmas carol) 139
Simoni, M. 60, 61
Sinfonia (Berio) (composition) 138
single-time dynamic visualization 87–8, *87*
Smalley, D. 2, 44, 53
Sommeil (Henry) (composition) 58
sonata (form) 70, 73
sonic causality 113–14
sonic narrative 110

Sonic Rube Goldberg Machine 76, 113
sonic stroke 33
sound design 35, 50, 55, 64, 65–7, 70, 72–3, 75–7, 141, 142, 171, 172
sound event 110
sound object 12, 36, 46, 57, 76, 113, 165
sound organization. *See* organized sound
soundscape music 59
soundscape recording 46, 47
source bonding 53, 110
source descriptors 90, 112, 144
Sousa, D. P. E. 156
spatial 39, 68, 69, 71–3, 75–8, 91, 106, 113, 137, 139, 146, 150–4, 157, 158, 164
　animation 77, 147, 168
　configurations 5, 115–16, 118, 119, 151, 156, 167
　cues 60, 134
　layers 66
　movement 36, 58, 72, 154, 162
　properties 160, 174
spatiotemporal organization 72–3
spatiotemporal universe 5, 123, 125, 129, 134, 150, 158–60
spectral 18, 28, 69, 75, 145–7, 149, 162, 164–5, 168
　cues 134
　properties 37, 144, 145, 160, 174
　qualities 72
spectrum 1, 16, 28, 43, 67, 71, 100, 136, 137, 146
spectral stretching 162, 165
speech 100
Spellbound (Hitchcock) (film) 20
Spinoza, B. 33, 131
staccato gesture 36, 72–4, 148, 158
stereo separation 116
story 129
stochastic 62, 63, 66, 75
Stockhausen, K. 27, 42, 71, 108, 116, 138, 173
Stravinsky, I. F. 138
stream segregation 146–7
Studio di Fonologia Musicale (Milan) 41
Subotnick, M. 13, 18, 38, 75, 136
Switched-On Bach (Carlos) (composition) 19, 20
symbol (Charles Sanders Peirce) 101

symbolic meaning, communication of 48
synthesis 44, 78, 159, 162, 174
　bird-vocalization 66
　frequency modulation (FM) 65, 66, 70
　granular 19, 65, 66, 76, 135, 136, 147
　pulsar 65
synthesizer 7, 16, 18–19, 43, 69, 70, 75–7

tape music 16
tape splicing 19
taxonomic listening 58
technological listening 158, 159
telephony, electrical 8
temporal 27, 28, 31, 39, 70, 74, 86, 87, 107, 129, 146–7, 152, 157, 160, 161
　configurations 5, 115–16, 118, 119, 149
　contour 164, 167
　flow 165–6
　organization 69, 72, 148
　properties 148, 160
　unfolding 36, 66, 72, 113, 139
Theremin 20
Thomas, J.-C. 58
Thomas, N. 102
Thousand Plateaus, A (Deleuze and Guattari) 33
timbre 9, 52, 66, 72, 108
time, sense of 157–8
time stretching 115
Tin Men and the Telephone 72
Toch, E. 9
tonal thread 73, 140, 161, 163, 164, 167, 168
total serialism 12, 40, 108
Touch (Subotnick) (composition) 75, 76, 136
Touche pas (Roads) (composition) 62, 64 n.1, 75–9, 115–16, 138, 148–9, 159
　affordances and affect in 135–6
trace 48, 49, 104, 107, 117–18, 129, 131
Trainor, L. J. 26
transistor 7
Trautonium 8
Trehub, S. 26
Truax, B. 39, 103, 142, 148
Tsang, C. D. 26
Tubby, K. 20
Turkish bağlama 28–9

turntablism 9
twelve-tone technique 12, 18

units
  event 99
  narrative 2, 5, 109–12, 118, 119
  structural 5, 71, 109, 121
unreality 2, 50, 70, 117. *See also* negation of reality
unqualified experience 132
unstructured musical sequences 148, 149
Ussachevsky, V. 138

vacuum tube 7, 8
Vaggione, H. 35, 38
VanDerveer, N. J. 100, 103
van Egmond, R. 53
Varèse, E. 10, 11, 27, 47–8
Västfjäll, D. 31, 143
Very Last Christmas, The (Tin Men and the Telephone) (album) 72
Vines, B. W. 108
virtual 2, 10, 15, 39, 44, 106, 115, 133, 166, 168
virtual reality (VR) 33, 172–3
visual arts 3, 121, 127
visual domain 104, 147, 150
visual imagery 31, 162, 169
visual media 20, 133, 172
visual representation 37, 102
vocabulary of music 30, 40, 173
vocal 138
  lines 155
  phrases 156
vocal learning 24
vocalization 24, 66, 70, 137, 154, 163

vocoding 67

Wagner, R. 138
Wallin, N. L. 26
Watson, C. S. 111
waveform 36, 58, 68–70, 73, 77, 142, 161
Ways of Worldmaking (Goodman) 122
Weale, R. 59
Weaver, M. 106
Weburn, A. 40
weighing in worldmaking 122, 129
Wen-chung, C. 11, 48
Westdeutscher Rundfunk (WDR) (Cologne) 40, 42, 67 n.2
Westerkamp, H. 59
white noise 67
Whittall, A. 17
Wireless Fantasy (Ussachevsky) (composition) 138
wire recorder 9
Wishart, T. 44, 150
Wish You Were Here (Pink Floyd) (album) 139
worldmaking 2, 5, 119, 121, 150, 173
  diegesis and 122
  (see also diegesis)
  ways of 122
world semantics 128
world version 122–5, 129, 166
Wurmfeld, S. 127

Xenakis, I. 13, 19, 108

Zagonel, B. 108
Zampronha, E. 108
Zappa, F. 61
Zattra, L. 63

Printed in the USA
CPSIA information can be obtained
at www.ICGtesting.com
LVHW010400011023
759717LV00002B/244